T4-AFY-758

Healing Appalachia

Healing Appalachia

Sustainable Living through Appropriate Technology

Al Fritsch and Paul Gallimore

The University Press of Kentucky

Publication of this volume was made possible in part by a grant from the National Endowment for the Humanities.

Scholarly publisher for the Commonwealth,
serving Bellarmine University, Berea College, Centre College of Kentucky, Eastern Kentucky University, The Filson Historical Society, Georgetown College, Kentucky Historical Society, Kentucky State University, Morehead State University, Murray State University, Northern Kentucky University, Transylvania University, University of Kentucky, University of Louisville, and Western Kentucky University.

Editorial and Sales Offices: The University Press of Kentucky
663 South Limestone Street, Lexington, Kentucky 40508-4008
www.kentuckypress.com

11 10 09 08 07 5 4 3 2 1

Diagrams by Mark Spencer.

Library of Congress Cataloging-in-Publication Data
Fritsch, Albert J.
Healing Appalachia : sustainable living through appropriate technology / Al Fritsch and Paul Gallimore.
p. cm.
Includes bibliographical references and index.
ISBN-13: 978-0-8131-2431-5 (hardcover : alk. paper)
ISBN-10: 0-8131-2431-X (hardcover : alk. paper)
ISBN-13: 978-0-8131-9177-5 (pbk. : alk. paper)
ISBN-10: 0-8131-9177-7 (pbk. : alk. paper)
1. Appropriate technology—Appalachian Region. 2. Sustainable living—Appalachian Region 3. Bioregionalism—Appalachian Region. 4. Environmental protection—Appalachian Region. 5. Green products. I. Gallimore, Paul, 1947– II. Title.
T21.5.A67F75 2007
728'.047—dc22
2007001530

This book is printed on acid-free recycled paper meeting the requirements of the American National Standard for Permanence in Paper for Printed Library Materials.

Manufactured in the United States of America.

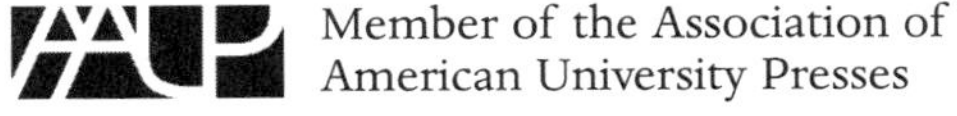

Contents

Illustrations

Acknowledgments

We would like to acknowledge all the folks in Appalachia who are living their lives in an Earth-friendly and sustainable manner. They are our hope and inspiration.

We would especially like to acknowledge all who helped make this book possible, especially Mary Davis and Robert Davis for editing and resource documentation, Mark Spencer for his diagrams and photography work, Warren Brunner for the use of photos from his Appalachian collection, and our Web site and Internet specialist, Janet Powell, for technical help. We deeply appreciate the critical remarks made by Michael Frome and John Nolt in their review of the draft copy. We thank these individuals for their quotations: Gary Anderson, Chris Ahrens, Joshua Bills, Deb Bledsoe, John Davis, Bob Fairchild, Pat Gallimore, Tod Garland, Richard Hotaling, Stan and Patti Jones, Joey Kesler, Jack Kieffer, Martha and Richard Lammers, Jeffrey Larkin, Jennifer Lindberg, Andy McDonald, Steve Mims, Arthur Purcell, George Rector, John Robbins, Mark Schimmoeller, Jerry Schumacher, Rhonda Sherman, Becky Simpson, Mark Spencer, Don Surrett, John Thomas, Carol Warren, Gene Wilhelm, and Syl Yunker. We thank the staff of Appalachia–Science in the Public Interest (ASPI) (especially Ben Perry and Martha Bond) for furnishing reference materials; Charlie Fritsch for technical assistance; and Ed Lammert for his garden photograph.

A number of people have been very generous with their time, energy, and creativity in support of the real work that underlies the foundation of this book. They have provided genuine inspiration, kindness, and compassion over the years.

I (AF) thank my parents, who lived the simple life described in this book, along with friends who have inspired me over the years, especially Jesuits Fred Miller, Bob Sears, and Jack Kieffer, and also Ralph Nader, Irene Dickinson, Richard Garascia, Michael Dewar, Wendell Berry, John and Sandra Freda, Pat Hoover, and distant relative E. F. Schumacher. Thanks are extended to the many volunteers and board and staff members at ASPI not already mentioned, who over the years made appropriate technology possible. Additional thanks are due to St. Elizabeth of Hungary Parish in Ravenna, Kentucky, for the use of physical facilities and to the Kentucky Jesuit Mission for financial assistance.

I (PG) offer infinite gratitude and appreciation to all my ancestors, and to all my teachers and healers: Helene, Robert, Lani, Malia, Pat, Sherando Joy, and Yanu Gallimore; and to Ted Williams, Sara Smith, Grandma Bertha Grove, Grandmother Redleaf, Tadodaho Leon Shenandoah, Joshu Sasaki Roshi, Gary Snyder, Jean MacGregor, Art Horn, Janie Westmoreland, Paula Bell, Walt Clark, Sara Martin, Dustin Cornelius, Greg Olson, Rick Maas, Wally Bowen, Chris Ahrens, Chris Shea, Clyde Hollifield, Debra and Joe Roberts, Brad Stanback, Paul and Inez Ghost Horse, Diana Osborne, Renate Sorenson, Aleah Wicks, Tony Martin, and Jerome Chambless.

And I offer gratitude and hope for all future generations, especially my granddaughter, Madison Clifton.

Finally, we wish to thank Joyce Harrison, Ann Malcolm, and all the folks at the University Press of Kentucky for their valuable assistance in seeing this book to completion. We would like to thank Cheryl Hoffman for her editing of this manuscript.

Introduction

A nation deprived of liberty may win it, a nation divided may reunite, but a nation whose natural resources are destroyed must inevitably pay the penalty of poverty, degradation and decay.

Gifford Pinchot, *The Fight for Conservation,* 1910

The Kentucky hills have enchanted me from my earliest youth.[1] I had to help with farm chores, including milking, from age six. Often during these years I would look out of the cow parlor to the east and see the rising sun over the wooded Appalachians in Lewis and Fleming counties. That sight of the sunrise over the hills is deeply ingrained in my memory. I came to love the Appalachian highlands and felt a call to always regard the valleys and coves of this part of America as home. With time I saw the devastation of these mountains by unsustainable forest practices. I observed the very shapes of our hills melt before the giant earthmoving equipment skinning away forest cover, exposing coal seams, and hauling away the black gold. Through this travail the hills called out for renewal, and this cry of the poor Earth brought me back home to Kentucky in 1977.

Several times I have visited the forested hills of Alsace (the Parc Naturel Régional des Vosges du Nord in northeastern France) and Schonau in the adjacent Pfalz region of Germany from where both my paternal and maternal ancestors came. My

1. In this introduction and throughout the book, all first-person singular references in the text are to Al Fritsch.

ancestors saw the sun rise each morning over forest-covered hills surprisingly similar to those in Appalachia. In fact, my grandfather for a short period was a caretaker of some of those Alsatian oak forests. Perhaps the love of these lands was why forebears on both sides of my family gravitated to Appalachia with so many other Alsatians: the Appalachian hills and the Ohio valley looked so much like the Vosges Mountains and the Rhine valley.

The love of land and people runs deep in our family history, and the impulse to enjoy, preserve, and renew these beautiful hills in both America and Europe is in my blood. Today in Europe coal mines are closing and forests are being properly managed. My relatives' villages are thriving principally by means of tourism throughout the year. Four bed-and-breakfast places exist in the tiny native Alsatian village of Obersteinbach next to our ancestral hamlet. The region has many motels, hiking trails, shrines, castles, historic sites, and beautiful roads. The success of Alsace is in part the inspiration for the book *Ecotourism in Appalachia: Marketing the Mountains.* The Appalachian region could welcome many visitors as well, and that is my dream; it could return to a condition similar to that of its European counterpart.

This work on Appalachian appropriate technology (suitable ways to achieve a higher quality of life while using fewer resources) is a natural sequel to the ecotourism work. In that book we defined ecotourism as a practice that enhances the environment and the resident population as well as furnishes a worthwhile experience for the visitor. These components cannot coexist in a region where inappropriate technologies (all energy from nonrenewable resources, all food imported from outside the bioregion, impaired water quality, improper waste disposal, etc.) are dominant. Instead, the technology and lifestyle of our people must be appropriate or suitable to the circumstances in which we find ourselves. Often the practices we advocate in this book are quite similar to traditional Appalachian ways. In many respects, this lifestyle is a return to our roots.

We refer often in this book to "Appalachia" and "Appalachian." A word of explanation is in order on this point. We have always preferred to leave the boundaries of our region a little undefined, as in *Ecotourism in Appalachia,* with a focus on the five central Appalachian states (West Virginia, western North Carolina, western Virginia, eastern Kentucky, and eastern Tennessee) along with southeastern Ohio. The federally defined region includes a northern tier covering parts of Maryland and New York and much of Pennsylvania. The southern end of the federally defined region includes parts of Georgia, South Carolina, Alabama, and Mississippi. From a geological perspective, we would have to go all the way and include New England and the Maritime Provinces of Canada. We think our original focus in our environmental work on central Appalachia best suits the purposes of this book.

But appropriate technology is a necessity for our planet as well as our country and the Appalachian region. We hope to offer a regional model of what the rest of the country and world could do and be. This is a place where good ecology and good economics could come together. If we are to attract a substantial number of tourists to Appalachia, our naturally beautiful region must divorce itself from its heavy dependence on resource-extractive industries such as coal mining and ecologically devastating timbering. Trees provide more jobs when they are left standing than when they are logged. The U.S. Forest Service reports that 74 percent of jobs provided by national forests come from recreation (and only 3 percent from timber). Keeping the forests intact will allow for a tourism that will fuel the economy of our region, furnish a livelihood for the people, and still preserve the unique environment of Appalachia. The challenge is to offer low-priced, people-friendly, and ecologically sound solutions to problems so that people and land may thrive together.

The story of Appalachian appropriate technology needs to be told. What is offered here is a summary of the hands-on experience from over a quarter of a century of work by two Appalachian demonstration centers as well as the testimony of knowledgeable and enthusiastic people who work within the thirty subject areas treated. We have described many times, in environmental resource assessments for some two hundred non-profit groups throughout North America, what we are presenting here.

The assessment program was initiated by the Appalachian Institute, a coalition of a half dozen public interest groups in central Appalachia under the National Science Foundation's Science for Citizens Program. That federal funding source was abruptly terminated in January 1981 under the then new Reagan administration. However, our own environmental assessment program continued to function by its grit and rugged determination. This program was envisioned as a way to promote solidly based Appalachian contributions to appropriate technology to the rest of the nation, while earning enough money to fund our Appalachian demonstration centers. For over a quarter of a century we have helped prove that Appalachia has more to export than just coal, timber, and a hardworking labor force. And ours is the only original Science for Citizens program still functioning.

Naturally, the written results of those environmental resource assessments belonged to the groups that entered into contracts with us and could not be directly transferred in report format to the general public. However, over time we have written and edited materials in the thirty subject areas, while leaving out proprietary information and the names of the specific assessed sites; here, we have focused on material generated at our demonstration centers in Appalachia and taken additional examples from nonassessed regional sites. To present results that are manageable by readers already burdened by too much information, we have culled the massive amount of data to present succinct descriptions, interspersed with charts, diagrams, and pertinent references for further study.

This book is meant for you, and our hope is to ease your hesitancy at matters of technological innovation by giving down-to-earth solutions to frequently raised problems. We believe it is fitting to bring this to you through an academic press devoted to advancing our Appalachian region. At the same time we will continue to furnish you with our technical publications, talks, local and regional demonstration events, the regional *Earthhealing* television show, and efforts to publicize results through fairs. This book is meant to be a centerpiece in an ongoing program to popularize alternative technology and to inspire individuals and communities to roll up their sleeves and begin the real work. Some technologies, such as more efficient automobiles and solar and wind energy applications, are already becoming quite popular. Others, such as cisterns and certain food preservation techniques, were employed by our grandparents and over time have fallen into disuse but can again be fruitful practices if modified. We do not expect you to use ideas from all thirty areas, but we are confident that some will fit your needs.

THE GLOBAL SITUATION

In 2001 the United Nations Environment Programme issued the report *Natural Selection: Evolving Choices for Renewable Energy Technology and Policy.* The report's authors argue that from $9 trillion to $15 trillion will be invested by 2020 in new global power projects. If a greater part of this massive investment were made in clean renewable energy technologies, the global economy would be more secure and more robust, and Earth much cleaner than in the twentieth century. Energy demands continue to rise, as they did through the 1990s, at about 2 percent a year. To continue to meet this increasing demand through traditional means (heavy reliance on nonrenewable energy sources) will increase pollution levels, kill 500,000 people a year, and cause, among other ailments, 4 million to 5 million cases of chronic bronchitis. These health-destroying and environmentally devastating energy production methods will continue to cause acidification of ecosystems, contaminate the soil and water, imperil biodiversity, and contribute to the catastrophes associated with global warming.

The United States has all the resources necessary to take a lead in cleaning up the damaged global environment and converting to renewable energy. With enough public awareness and pressure, perhaps some of the immense—more than $400 billion—annual military expenditure required in part to protect nonrenewable energy sources could be transferred to development of solar and wind energy sources. The need to review our national consumption patterns in many lifestyle areas has never been greater, and America should take a lead. How much longer can we afford to wait?

For better or worse, Appalachia is squarely in the middle of this global struggle between nonrenewable (coal and oil) and renewable energy sources (solar and wind). Being near the heart of the United States, Appalachia has become a showcase of destructive resource-extraction practices whose extreme damage to our land, air, water, and people is becoming evident. Kentucky has the second-highest rate of respiratory death and illness due to power plant emissions, and other Appalachian states crowd right around it. More than cleaner coal, our people desperately need renewable energy substitutes that are healthier and more environmentally sound. Appalachians need not wait for the nation; they can lead the way. That has been our dream.

Appropriate technology is a global, national, and regional enterprise and involves both giving and receiving. Are we willing to modernize traditional practices such as food-growing techniques and use of native building materials—and share these practices with others, especially in so-called developing countries? Are we willing to adapt a wide variety of local and other appropriate technologies and see them as a continuation and celebration of our past traditional practices? Or must all of us accept the recent discontinuity with our cultural history, which has caused a younger generation to forsake its roots and become fast-food junkies and uncritical consumers of a variety of unneeded products?

When these basic questions arose in the 1970s, I regarded their answers as a real ongoing puzzle—the kind that you might also like to solve. I returned from Washington, D.C., to Kentucky in 1977 and was anxious to see whether suggestions proposed in the book *99 Ways to a Simple Lifestyle* could be successfully demonstrated at the grassroots level—and on my home turf. In fact, the appropriate technology demonstration phase of this work initiated then has been quite successful, as we describe in later chapters. It was a stroke of luck that we chose to demonstrate in our home bioregion of Appalachia; a majority of the demonstrations would have had so many regulatory barriers on the eastern seaboard that we might have been discouraged from working on them. Appalachia has proved a more tolerant environment.

CHALLENGES AHEAD

The widespread popularization of these proven appropriate-technology techniques is more problematic and requires a concerted effort by everyone from energy advocates to auto dealers, from nursery salespersons to agricultural county agents, from schoolteachers to clergy. Our testing programs at both ASPI at Mount Vernon, Kentucky, and the Long Branch Environmental Education Center at Leicester, North Carolina, have covered the subject areas and have involved feedback from staff and interested users. In our modern American culture lower-income folks have been panicked into following practices of the more affluent in the implicit belief that a more sophisticated technology is a better one. But some appropriate technology practices such as renewable energy applications and organic food production are certainly gaining acceptance among all income groups both in the region and elsewhere. The challenge ahead is for Appalachia to take the lead in popularizing proven appropriate technologies.

Cleaner renewable energy sources are available and ready to go. The United Nations report cited above states that after starting from a small base in the 1970s, biomass, geothermal, solar, small-scale hydropower, and wind technologies have grown proportionately faster than any other technologies for supplying electricity. Wind energy, the world's fastest-growing energy source, has far exceeded the most optimistic 1990 projections. Its price has dropped to one-seventh its price in 1990, and wind is now becoming competitive with fossil-fuel technologies, even with their governmental subsidies. Renewable energy sources are becoming more attractive due to the increasing scarcity and inherent pollution of nonrenewables. The report cited above states: "It is increasingly true that there are no technical, financial or economic reasons why the nations of the world cannot enjoy the benefits of a high level of energy services and a better environment. It is simply a question of making the right choices."

One may ask, if the report is restating what many appropriate technologists have said for decades, why is it taking so long to convert to renewable forms of energy? The answer rests in the power of the nonrenewable energy companies exercised through subsidies and tax breaks. The oil, gas, coal, and nuclear conglomerates are large and powerful, and they strongly influence governments and will not relinquish their control. However, some companies are breaking ranks and seeing value in the production of renewable energy. As long as favored energy and other practices continue, economic advantages are far too high for many companies to be willing to risk their market share and influence. But economics is not the total picture, and that fact triggers a second look at what is truly profitable. When health and environmental factors are included in the equation, the current dominance of nonenewable energy does not make sense. Environmental degradation, deteriorating health conditions, and pollution are not suitable for future generations.

Over half the technology applications mentioned in this book deal with renewable energy, an area in which ASPI and Long Branch have played prominent roles (e.g., the initiation of net metering of solar energy in Kentucky and the development of the solar compost toilet in North Carolina). We still fight an uphill battle to gain widespread acceptance within our region. Appalachian people are just as prone as other Americans to be influenced by commercial promotional campaigns. Our people also tend to choose larger numbers of mobile and manufactured homes, fast foods (and accompanying heavy consumption of soft drinks), and gas-guzzling vehicles. However, it is not our intent to critique current practices, for there is ample literature on the subject; rather, it is better to promote appropriate practices that are affordable, people-friendly, community enhancing, and environmentally benign or beneficial.

The time is right for Appalachia to take the lead in appropriate technology. Our region is known for its resourcefulness, self-sustainability, and the ability to make do with very little. Ours is a region of great promise, a potential showcase of conversion to renewable energy and practices that use fewer resources. Regulations, some of which could perhaps promote appropriate technology, can also hinder the research and development of such appropriate technologies as innovative low-cost shelters and space-heating applications, though they are not nearly as restrictive in our region as they are in other regions. From the school of hard knocks we Appalachians have learned to take matters into our own hands; our precious dollars can be recirculated and recycled at home; we can see the fruit of our labor; we can make critical decisions about the life of the local community. Currently, extractive industries—most often under outside control—cast a dark shadow over the region's towns and villages through increased pollution, environmental degradation, and resource depletion.

However, land and people are intrinsically tied together, and when the land hurts, so do the people. Damaged land means damaged communities. With high rates of drug and physical abuse in our region, a sense of despair and powerlessness creeps across our land. Appropriate technology is not the entire answer but it is a beginning. As it gains acceptance we will see out-migration slow or cease, decent employment return, the hills heal, and our regional self-esteem gradually improve. Appropriate technology stimulates these improvements because its practices are people- and community-friendly, and that all adds to a growing sense of hope.

CRITERIA FOR SELECTING APPROPRIATE TECHNOLOGIES

The term "technology" is often used today to refer to computerized devices and processes, but it obviously has a far broader meaning. Technologies go back at least in very simple forms to the dawn of human existence. Technology developed slowly through the millennia and has involved a wide variety of activities: production of fire; flint knapping; basketry; making pottery; cloth weaving; plant and animal breeding; use of wheel, sled, and animal harness; iron forging and copper smelting; various designs of plows and tillers; the windmill and waterwheel; and on and on. Older tried-and-true methods have given way to improvements over the centuries, and in some, but not all, cases these changes have enhanced the appropriateness of technology at a given time and place. Without forgetting the achievements of civilizations, the challenge today is to incorporate the new in such a manner that people's lives, both individually and collectively, are improved while the Earth itself is not irretrievably harmed. For a modern technology to be appropriate, it should also be:

Affordable. Technologies that people cannot afford, or that cause them to incur long-term debt or mortgages, are not appropriate. Such indebtedness does not lend itself to a high quality of life. Often the pressure of keeping up with others influences many to devote time and resources to modern expensive technologies (fuel-guzzling vehicles or energy-inefficient housing) that demand more and more attention and working hours to purchase and maintain. Affordable means a higher quality of life within the means of average citizens while allowing time for leisure and enjoyment.

Earth-friendly. Refraining from polluting the air, water or land and depleting precious resources is at the heart of all appropriate or proportionate human activity, including the tools to achieve the activity. If we damage an environment by joyriding on an off-road vehicle in sensitive terrain, then this is recreation turned to unconsciously hostile action and thoughtless ecological destruction. Ultimately, bad ecological practice harms natural and human communities as well as the perpetrator. Those creating appropriate technologies should aspire to meet basic human needs while preserving and protecting the life-support systems and precious biological diversity of the entire planet.

Community-enhancing. An appropriate technology is not so exclusive and automatic that ordinary people are removed from its care and maintenance. A good technology is a community-building or cooperative venture involving a number of inhabitants. Certainly all communities will not benefit equally, for communities have different needs. However, an appropriate technology under ordinary circumstances will bring together rather than divide, heal rather than fracture, and help build up the spirit of the total community through mutual benefits from its use. A complex technology that is inherently elitist is not community enhancing.

People-friendly. An appropriate technology can be learned and maintained by people with ordinary skills. Many of us know people who are terrified by fine-print instructions on a complex device or by being required to take courses to learn to use a new piece of software. A few take the increasing demands in stride and master the technology, sometimes arrogantly assuming that everyone else should follow their example. Technologies should not be so oppressive that

> those of us with moderate, but not great, technical dexterity are left out. A technology should be enjoyable and spiritually enhancing to the individual user.

Many of the thirty technologies mentioned here are gender neutral or slightly favor traditional Appalachian male roles (solar applications, and wind, wood, and silvicultural practices). Appalachian women often still have the role of traditional homemaker and gardener. In recent years gardening is being taken up by male retirees and lawn care is being done by men of all ages; both men and women are showing interest in other outdoor aspects of the domestic scene (ponds, forests, wildlife, and on-site construction). Since we find both men and women performing the various functions (with the exception of food preparation and some wildcrafting), it appears that the gender barriers are breaking down even in Appalachia. Joint participation can occur in various ways: in yurt, cordwood, pressed earth, and cabin construction through barn-raising events; in joint care of a solar greenhouse; in sharing of the solar electric car on short trips; in annual tree-planting events for all the family; and in cooperative efforts at energy and water conservation.

Need for Criteria

Selecting examples of appropriate technology is more difficult than one might first think. A variety of tools and activities may fit some of the characteristics and yet not be applicable to all times, places, or communities. In preparing environmental assessment reports and recommendations we have discovered that some examples of appropriate technology are generally practical for most people, sites, and cultural or personal resources, and others are too inherently difficult for particular organizations or individuals.

Not all practices termed "ecological" or "green" are good for everyone who hears about them. Often people are so enthusiastic that they will try too much and do few things well; sometimes they will try too little for fear of failure. Suggesting practices according to available assessed resources over a specific time period (e.g., ten years) is at the heart of good environmental resource assessment. Our reports have always directed clients to areas where we were confident they could demonstrate an effective outcome, and we have sought to detour them from areas that would most likely not provide some beneficial results. We would recommend creating a wildscape if a lawn area would require excessive care. On the other hand, we would not recommend a wind generator if wind power potential is very erratic at a given location.

In selecting technologies, one must look for the good. All of us are limited as to resources such as finances, time, and talents. To paraphrase the Good Book, while the spirit to be Earth-friendly is willing, the flesh is weak (inclined to take on too much or too little). It is best to emphasize limited but highly appropriate applications, giving details as to the relative importance of applications, how much time should be spent, and in what order a particular technological innovation should be implemented. So often an analytical critique of options is omitted from assessments of resources, and service technologists will do whatever a client desires, leaving the assessment to the individuals rather than outside and objective sources. Outsiders will be happy to do audits of resources (which we believe those in control should do themselves) and then leave assessments (which we believe should be done by outsiders) to the target group. We are convinced after two decades of work in this field that a book on just such critical analysis is needed today.

Good for All

Some general technological applications, especially in areas dealing with providing basic human needs, are good for all people in some fashion. For instance, better forms of communication or of making information available are for the good of the entire human race, although not all individuals may have access to them and may need community assistance (see Postscript).

The four areas in which human beings have the greatest need are food, water, shelter, and fuel (all bulk goods or things that are difficult to transport great distances). Good ecological practice and lessons from the beginnings of civilization have evolved some basic practices of hunting, gathering, or growing food locally, using local water supplies without contaminating them, making shelters with local materials, and finding fuel from sources close at hand. In the same manner, eliminating waste through recycling and composting at the local level preserves land, air, and water quality. The immense environmental cost of transporting bulk materials from or to distant places, along with the complex system for securing these places, often outweighs any advantages of acquiring distant resources. If our nation continues to spend immense amounts of money to defend distant fuel supplies or other resources, we will simply go broke in a few decades.

A distinction can be made between elite and democratic technologies. Communicating a technical idea or transporting a lightweight, seldom-used device over greater distances (a hoe used for cultivation or a house design) can help create a higher quality of life for everyone who participates in the exchange. We will both share technologies that have proved to be effective and continue to look for appropriate technological ideas and innovations from distant peoples. We can ill afford to pretend that people in the "developed" world, or the more affluent members of the human race, have all the answers. In fact, it is a mark of maturity to seek out simple-living folks in every culture and find out how they keep themselves healthy and their ecosystems intact through their ingenuity in using simple tools and practices.

Reality check: Some may believe that Western civilization with its Greek architecture, Egyptian pyramids, Roman road systems, and modern skyscrapers marks the epitome of what human beings can achieve. Yet the Athabascan peoples of the Pacific Northwest coast of North America have lived there for over ten thousand years, and, until European contact, the ecosystem there was fully intact—no severely degraded habitats, no extinct species, and an abundance of fresh water, wild foods, and wood resources for shelters and fuel. So will the real civilization please step forward . . .

Good for Some

Some technological applications are practical for people in a given place or culture or for those with certain personal talents and resources, but not for everyone, everywhere. A building design may work quite well where soil conditions, topography, humidity, and available native materials allow, but that same design should not be attempted in other areas where difficulties may occur (this is *bioregional specificity*). Cultural differences may limit the use of a device because of food preparation, shelter requirements, or travel needs that render the practice quite suitable in some cultural circumstances but not in others *(cultural specificity)*. One person may find it quite appropriate to undertake a given practice for the good of the whole community, a practice that keeps him or her employed and permits recycling certain materials *(personal specificity)*, but this would not necessarily mean that everyone even within the specific locality could undertake this practice for lack of raw materials, markets, or ease in obtaining desired results.

Good for None

We do not want to be burdened by proving the inappropriateness of specific technologies, for that would most definitely go beyond the limits of this book. If a practice is inherently dangerous to users, neighbors, the Earth, or future generations, or can only be used by a few highly trained experts, or will deplete resources in its applications, or will result in concentration of economic power in the hands of a few to the detriment of the entire community, it should be omitted from consideration. Nuclear power technologies as applied to electric generation come to mind. *Critical Hour,* a book I coauthored with Mary Davis and Art Purcell, lists a number of reasons nuclear power for electric generation is an inappropriate technology: long-term waste disposal problems, lack of safety of generating equipment as evidenced by accidents at Three Mile Island (near Middletown, Pennsylvania) in 1979 and Chernobyl (in the former USSR) in 1986, hazards associated with mining and processing of uranium source materials, inability to ensure safe transport of nuclear materials, terrorist threats at many power plant sites, and the high cost of total electric generation including governmental processing subsidies.

The absence of discussion of a particular example does not necessarily mean the technology is inappropriate. Our listing is simply directed to what appears to us to be more basic and widely applicable. The treatment is not exhaustive, especially when it comes to a host of smaller devices: small hand tools, personal items and cosmetics, household utensils, electric appliances, computer technologies and applications, communication devices, and recreational equipment and practices. The list could go on and on. Certain conveniences are luxuries for the able-bodied but necessities for the physically challenged; others are too recent to reveal hidden hazards; still others are part of an enticing consumer culture, which verges on creating addictive behavior disorders. All technologies are to be used with much care and mindfulness—qualities lacking in a world saturated with gadgets, gimmicks, and advertisements. Great care and mindfulness are general principles that apply to the use of all small devices. Sometimes too much technology can overwhelm our sanity and degrade our quality of life.

The use of extremely heavy sport-utility vehicles (SUVs), is deemed inappropriate because of high fuel consumption, even though some brands may provide some degree of safety in collisions. It is also quite possible that what appears inappropriate today may later be adapted or modified so that the harm done is highly reduced or eliminated.

EXAMPLES

Let's examine some specific technologies that fall into the categories of good for all, good for some, and good for none.

Good for All

The thirty ideas treated in this book are meant to be generally practical for all. Technologies such as solar energy and wind power have the potential for global application, but obviously some regions are better suited to their use than others. But the total mix of renewable fuel sources (solar, wind, wood, microhydropower, and biogas) would be to the net benefit of all people and the environment. Land reclamation may not be needed in some pristine subregions, but these unfortunately become fewer with the passing years. However, the book deals with aspects of global human living that require bulk products: fuel (electric energy, energy efficiency, and transportation), food, shelter, and waste- and water-related topics. Technologies in two areas, forests and land, are generally regionally specific by reason of their subject matter (deserts no longer have forests). However, more broadly, all land resources must be preserved, protected, seen as limited and fragile, used with care, enhanced through ecological restoration, and passed on to future generations in as good or better condition than that in which they were received.

Regional specificity. Within the general subject areas, attention to appropriate technology is given to examples that permit use over the broadest areas of a specific bioregion. Not all examples may apply here within a single bioregion. For instance, in our temperate Appalachian climate shade trees and windbreaks are of immense value. However, a particular condominium may have no place to plant additional trees even though urban forestry is of immense importance throughout the region. Also some technologies (e.g., cordwood construction) are meant to apply to this targeted region (eastern North America) where forest resources are close at hand. Someone building in Key West, Florida, may not find this helpful except in its building principles. In nonforested areas, it would be appropriate to look for readily available products such as stone or earth.

Cultural specificity. Solar cookers, which are good for many, may not be the best choice for people who eat mainly fried foods whose preparation requires intense heat. But most populations have some foods that could be baked or cooked in solar ovens/cookers. Thus the applicability is more limited in some cultures than in others because of the way food is prepared. And some groups may regard individual practices as outside their religious or cultural traditions.

Personal specificity. Choosing what technology, or technologies, to adopt is perhaps more difficult than one might realize. People know whether they can undertake to use a specific technology based on their limited talents and physical condition. Saying no to personal use of such technologies as bikes, compost toilets, or solar greenhouses may have a good basis that all respect. The difficulty arises when, for example, a person decides on a passive solar design, only to realize after the shelter is finished that an attached solar greenhouse might have been a better choice, one that would have provided more useful horticultural space and an equal amount of solar heating during the winter months. While there is a certain relativism to such judgments, all are subject to critical analysis by the person and the larger community.

Good for Some

Some appropriate technologies could meet the four criteria mentioned above—that is, they could be affordable, Earth-friendly, community enhancing, and people-friendly—and yet not be recommended for people in the eastern United States (regional specificity), nor be part of present-day American culture because of practical considerations, nor be generally applicable for individuals due to limited resources, markets, current product use, or other circumstances. We give three examples here of what could be termed "conditionally appropriate technology" that would nevertheless not be applicable to our audience within the bioregion targeted by this book.

Bioregional specificity. Examples of technology appropriate for one region being misapplied in another are nowhere more apparent than in the section on shelter. Part of the problem is the standardization of home building in this country without regard to ecological and safety aspects. What is good for Alaska is sometimes alleged to be good for Florida. The importance of regional specificity is not often understood even by those who consider themselves "green." They sometimes unthinkingly adopt what is appropriate in another region with no regard to local climate conditions or the basic ecological principle of using locally available materials. Straw-bale housing offers an excellent example of bioregional specificity.

With the advent of "modern" straw-baling equipment in the late nineteenth century, resourceful farmers in the Sand Hills of western Nebraska created well-insulated houses on the Great Plains. They selected their building materials locally and chose designs for performance, availability, and durability in that bioregion. The materials were close at hand; the climate was suitably dry, and the mean relative humidity was quite low. The straw-bale shelter was and still is appropriate there, as it is across the West, except in the temperate rain forest in the Pacific Northwest. It is particularly appropriate in central California, where highly durable rice straw is often burned as a waste material, thus contributing to air pollution.

An examination of a mean relative humidity map of this country quickly reveals that this nation has varied humidity. If potential builders would respect the wisdom of local people and ask about using straw as a building material in the eastern United States, they would get the response about straw-bale housing given to me by an old farmer: "Are they crazy?" What this farmer and others close to the land know is that in our Appalachian climate straw becomes moldy when stored for a few years even in a dry barn. The straw-bale structure presumes low humidity, for the straw bale walls are meant to breathe, and this air can add moisture to the straw. In zones where the relative humidity is 66 to 75 percent, it is possible for the moisture content of the straw to creep up to 25 percent or higher.

It is uncertain whether the dangerous *Stachybotrys atra* mold, which causes the disease known as farmer's lung, will grow in the eastern climate using the well-known growth medium of straw. Some easterners who live in straw-bale dwellings are already making complaints at this time. What about infants and the very young who can't easily express their discomfort? Is this so-called appropriate dwelling becoming an example of moldy indoor housing, the twenty-first-century equivalent of asbestos contamination?

In more than half of the eastern United States, straw is not easily procured—but the same could be said of logs. Is it worthwhile importing straw bales, which are 24 inches thick, when every critical inch of walls requires that much more floor space and roofing? Is it worth following a national fad when people could be harmed in the process? Importing a nonnative material is less ecologically desirable than importing far more efficient and more spatially economical recycled cellulose or other forms of insulation that cause fewer potential health problems. (For more information on straw-bale housing, see ASPI Technical Paper 70, *Straw Bale Housing* [2003]).

Cultural specificity. Methane, the major component of natural gas, is well known as a relatively clean energy source, but while global resources are still quite substantial, those that are available for consumption by an energy-hungry population are limited. Liquefying natural gas from distant sources and transporting it in ocean vessels is dangerous, even when such ships are docked away from populated areas. But methane is also found closer to home. Incomplete anaerobic decomposition, as in wetlands or landfills, results in some methane generation. However, aerobic decomposition, as in a composting toilet with good airflow, leads to conversion to nutrient-rich, pathogen-free soil amendments, carbon dioxide, and water, and no methane is generated (see chapter 26). Methane is part of the normal natural cycles of life, and so it is tempting to capture it wherever possible and burn it as a fuel.

Human and animal waste can be collected and processed through anaerobic digestion of recycled wastes. In some countries, such as India and China, both residential and farm wastes are recycled; in other instances, only animal wastes are used. Residential-scale biogas applications are harder, or impossible, to find in the United States, where natural gas has generally been relatively cheap, although prices have recently begun to soar. Methane gas is contaminated and of lower fuel value, manure is not easily collected, and the labor-intensity of such processing is often beyond the normal course of activity. Where fuel is in short supply and manure is available in higher concentrations, such as in a feedlot or chicken house near a residence or place of business, there is a greater incentive to digest the wastes into methane. In moderate- to large-sized farms with confined livestock and operating liquid-manure handling systems, the possibility of converting wastes to methane is greater. Certainly odors remain a contentious issue.

Culturally, even with unforeseen incentives, it is extremely doubtful that Americans will choose to make methane from their waste products. This is especially true since the amount of fuel produced per household is relatively negligible and the wastes collected could be more easily recycled via a composting toilet and returned as a soil amendment to the garden. Very few if any Americans could be persuaded to adopt this methane-generating practice in or near their homes. It would be a cultural stretch for many to accept household-scale biogas digesters.

Having said this, what about medium-scale or community methane digestion? During the 1970s, some dairies and hog farms received grants from the federal government to develop methane collection systems on a moderate scale. Over time such medium-range systems proved costly and difficult to operate without subsidies; they require maintenance by farmers who are already trying to keep marginal operations going and who have only limited labor available. In addition, the relatively small amount of methane generated makes the economics difficult to justify.

In contrast, previously existing municipal solid waste landfills built according to U.S. Environmental Protection Agency specifications have generated considerable amounts of methane gas, which, while somewhat impure, can be processed to help generate electricity. EPA estimates that landfills could produce enough electricity to power 1 million homes nationwide. With conservation and efficiency improvements and renewable energy systems installed at the residential level, perhaps ten times the number of homes could benefit from methane-generated electricity. About five hundred landfill operations are currently being tapped for methane, but in many cases such wasteful disposal practices are used to eviscerate source-reduction efforts, reuse, repair, recycling, and composting opportunities. However, to use all resources mindfully, existing landfills need not flare off the dangerous accumulations of methane. Since we are already burdened by too many landfills, it is far better to use the methane they generate to produce electricity. With many communities already setting their sights on zero waste, we have hopefully evolved beyond the need to site new landfills using the feeble excuse that they would be acceptable for methane generation.

On a larger scale, methane is currently collected in coal mines, using a waste by-product of the mining operation to produce a fuel that is less polluting than coal. This may be a proper industrial application, but it goes beyond the limits of this book, which concentrates on appropriate technology practices for individuals or communities.

A second large-scale methane generation source could be large-scale animal feedlots. This may be analogous to organizing a blackberry cooperative using briars growing on strip-mined lands. Large feedlots are found especially in the Midwest and on the Great Plains. Many environmentalists consider them abominations because of the noxious odors, potential surface- and groundwater contamination, and the poor treatment of livestock. These environmentalists think that such feedlots should be eliminated even if some reuse of manure is possible (with proper carbon-nitrogen balancing it can be recycled as compost). These opponents tell us that grass-fed, free-range livestock produce healthier and higher-quality animal products. But the feedlots could take advantage of economies of scale in using the methane; given the circumstances, methane generation may prove better than wasting the material. Shifting our meat production to free-range, antibiotic-free, non–bovine growth hormone (BGH) livestock would diminish methane-generation possibilities. However, under current conditions, composting the manure would still be a beneficial reuse.

Large-scale biodigestion is another example of something that is culturally inappropriate on a small scale but that could be appropriate on a large scale. Much discarded organic material generated in wood-, food-, feed-, and fiber-processing facilities can provide feedstock for biodigestion. Where the biogas has an effective end use, perhaps both as process heat and cogeneration for electric production, and the residue can also be used as an organic soil amendment, it may be quite suitable to turn these wastes into a usable fuel.

Personal specificity. Fuels may come from a variety of agricultural products or by-products, and the popularity of these alternative fuels is increasing as gasoline and diesel fuel prices rise, taking a larger bite out of family and institutional budgets. As fuel prices increase, there is a noticeable effort to take combustible waste materials or virgin materials with limited markets and use these as fuels in our traditional internal combustion engines. A number of questions arise.

What about using waste materials from industrial deep fryers in potato-processing plants, snack-food factories, and restaurants for fuel for space heating or for running vehicles? Certainly disposal of waste vegetable oil is bothersome for restaurants, and green technologists who collect, filter, and use the material as a biofuel for vehicles in their communities are to be commended. They save money, recycle a waste commodity, and use a cleaner-burning fuel than petrodiesel. While this small-scale practice may be personally appropriate, it is especially so when the waste oil is destined for disposal in a landfill or burning in an incinerator. Since the use of waste vegetable oil for animal feeds has been banned in Europe since 2003, the question must be raised why the United States has neglected to address this issue. The highest beneficial use or reuse of a resource—beyond just a technological or resource management issue—must also become an ethical issue. Wasting is unethical.

In some areas, several factors could make some of these technologies practical for individuals if not for larger-scale use. These factors include the lack of waste vegetable oil recycling centers; degree of contamination of the waste oil; insufficient quantity of materials in a given community; and willingness, or lack thereof, of the fuel user to spend limited time and resources recycling materials. The use of biofuels can be an important strategy for transitioning to the use of solar electric vehicles powered through solar charging stations (see chapter 24).

However, the pedagogical value of recycling is crucially important even when the practice has limitations. Isn't it ethically important to count the educational value of earnestly striving to reuse things well, even when the reuse is not perfect or is only of marginal economic value? Whether something is appropriate for an individual cannot be judged categorically from a distance but must be determined by the judgments of individual technologists in the context of specific communities.

Another alternative fuel that raises questions is ethanol. Used for years to enhance octane ratings, ethanol can be made from crops such as corn, whose cultivation requires the consumption of a great deal of petroleum products, as well as from cellulosic feedstocks, biomass wastes, native fast-growing plants like switchgrass, and short-rotation woody crops like poplar trees. Because recent legislation allows generous federal subsidies for ethanol generation as an alternative fuel source, we need a definitive analysis of total resource gain and loss, including energy output gained as well as soil depletion and use of nonrenewable fuels for growing, gathering, and processing the ethanol.

Recently, a major controversy has erupted between David Pimentel of Cornell and Tad W. Patzek of Berkeley on one side and a sizable contingent of the agricultural scientific community on the other. Pimentel and Patzek reported that producing biofuel requires more energy than it supplies in the form of ethanol. They found, for example, that producing ethanol from corn requires 29 percent more fossil energy than it generates, and ethanol from soybeans requires 27 percent more energy to produce than it provides.[2] Should valuable agricultural land be diverted from food production for limited quantities of fuel for internal combustion engines? See the discussion of transportation (chapter 24) for other alternatives.

In Appalachia, only wood waste products are major biofuel sources, since large-scale agricultural operations are somewhat limited. However, some biological waste products may be areas of interest in this region, given recent lucrative incentives. No

2. "Biofuels: Too Little for Too Much," *Earth Island Journal* 20, no. 4 (Winter 2006): 9; available at http://www.earthisland.org/eijournal/new_articles.cfm?articleID=1007&journalID=85.

one argues against using biomass (wood) to produce thermal energy for heating homes. Our region's limited supply of arable land would be better used to grow vegetables and other edible produce. Thus to some degree the controversy is somewhat academic for Appalachia.

STRUCTURE OF THIS BOOK

The following chapters discuss thirty friendly technologies that meet our criteria as appropriate for Appalachia. The chapters cover the following topics: electric energy (chapters 1–3), energy efficiency and conservation (chapters 4–6), food (chapters 7–11), forest (chapters 12–15), land (chapters 16–18), shelter (chapters 19–23), transportation (chapter 24), waste (chapters 25–27), and water (chapters 28–30). One can quibble about the way we have chosen to cut the pie here; certainly some applications would fit easily within more than one category. We have tried to assist the reader by providing ample cross-references.

CHAPTER 1

Solar Photovoltaics

I have been installing solar electric systems in Kentucky and Tennessee for over a dozen years now, from small cabin systems to large house systems. Many of my clients have environmental concerns as a major motivation. Photovoltaic panels are the cleanest electricity generators that can be used almost anywhere. They offer a practical alternative to coal-generated electricity and the ravages associated with coal mining. My home's electricity comes from solar electric panels and I encourage and assist others to do the same.

Bob Fairchild, director, Eastern Kentucky
Appropriate Technologies, Dreyfus, Kentucky

Many applications of photovoltaics (PVs) are touched upon in this chapter: domestic electricity, energy-conserving lighting, PV accessories such as water pumps, and "net metering," a procedure for integrating solar-generated energy into an electric grid system, thus reducing the cost of self-contained solar systems. Other chapters deal with such topics as passive solar heating of residences, greenhouses and other seasonal extenders, heating water and cooking, and solar-charged electric cars.

ASPI (Appalachia–Science in the Public Interest) has three separate PV systems, one on its solar house, one on the nature center, and one system for the office. The office system was intended to be simply a solar charging station for the ASPI electric car, but we soon realized that it could also furnish electricity to run the office. This utility intertied system allows excess solar-

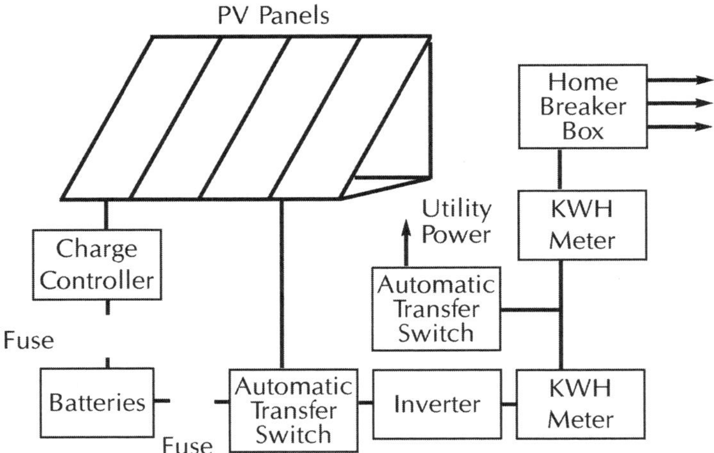

Fig. 1.1. Utility intertied system. A system such as this makes net metering possible.

generated energy to flow into the local utility company's grid (see figure 1.1).

The importance of electricity in the modern world cannot be overemphasized—and the massive Northeast blackout in the summer of 2003 underscored that fact. But must electricity be generated by environmentally flawed fossil-fueled or nuclear-fueled power plants in a world beset with global warming partly caused by power plant emissions and with unsolved nuclear waste problems? Solar, wind energy, and microhydropower systems offer clean, renewable sources of electricity. These technologies can be used at the residential level, but at some financial cost. Getting one's power from the sun, collecting rainwater, and growing one's food in greenhouses all combine to make the dream of a "living ark" come true, and this model can provide the springboard for achieving community interdependence.

Two movements seem to be occurring in the complex PV field: the decentralized generation of electricity through PV units and the centralized pooling of surplus energy from these units so it can be transferred to places in need. Decentralizing some generating sources while retaining transmission lines and utility grids for the welfare of a larger community challenges basic concepts of appropriate technology as self-sustaining independent systems apart from the larger world community. Solar PV can offer independence from the electric grid, but it becomes more economical when interconnected with the modern world. Perhaps we can join the best aspects of self-sufficiency and interdependence in a balanced way.

DOMESTIC ELECTRICITY

The potential for renewable sources of electricity at relatively low cost is at our own back doors and roofs.

AstroPower of Delaware was for years a leading producer of solar cells and modules made totally in the United States. They offered single-crystal 60-, 75-, and 120-watt modules with thirty-six series-connected cells and tempered glass glazing with aluminum frame and weather-tight junction box. The largest module came with a twenty-year warranty. Hopes were running high that these larger workhorse types of PV generating modules would become even more affordable in the near future when the company faced financial problems, was delisted from the NASDAQ in 2003, and filed for bankruptcy. AstroPower was purchased in 2004 by General Electric (GE Energy), which is making a commitment to renewable energy following its purchase of Enron Wind in 2002.

Solar shingles are also available. Recently attention has focused on companies such as Uni-Solar (a wholly owned subsidiary of Energy Conversion Devices, Inc.) that are producing PV roofing shingles that can also serve as a weather-resistant roof covering. In 2006 both solar shingles and PV panels are commercially available for about $4 or $5 a watt. It is anticipated that prices will decline for both. Economies of scale ought to apply to both technologies in the near future as solar growth in 2004 was up by 15 percent. These shingles come in 7 by 1 foot pieces and are fastened in place with common roofing screws and/or nails over conventional roof decking. Lead connection wires are on the back side, ensuring that electric connections are made in an attic (vaulted or cathedral ceilings would present some difficulties).

Other kinds of thin-film PV laminates that can be bonded directly onto standing-seam roofing panels come in 16-inch-wide peel-and-stick rolls and can be cut to length. Unbreakable thin-film PV is produced using amorphous silicon, encapsulated in Teflon and other polymers. In contrast to crystalline PV modules, there is no need for specially designed racks of heavy, expensive glazing. These thin-film sheets demonstrate improved performance in high temperatures and in partly shaded conditions. They also require one one-hundredth the silicon, which indicates that thin-film PV should become less expensive than crystalline PV as production capacity expands over the next several years.

Today, PV systems are found in various parts of the country, especially in the sunny Southwest. The British estimate that the entire UK's electric supply could be furnished if only about half of current roofing used PV shingles. The same could be said of Appalachia. Some questions await the experience coming from long-term use; they include the ease of repair, vulnerability to damage from rocks or storms, and the expertise needed to install and maintain these solar shingles. However, conventional PV modules have had over thirty years of operational testing in real-world conditions and have performed with greater-than-anticipated efficiency and durability.

Although much attention is given to the modules, many other aspects of the PV system demand expertise in installation. Besides deep-cycle batteries, the system must be equipped with circuit breaker disconnects for safety as defined by the 1999 National Electric Code. If storage batteries are part of the system, they need charge controllers to keep batteries from overcharging. The system could be set up to divert excess electricity to water- or air-heating elements. Owner/operators need a monitoring meter to stay informed of their system's status. An inverter is needed to convert low-voltage direct current (DC) to the 120-volt alternating current (AC) used by typical American household appliances and office equipment.

Many of these functions can now be performed by the easy-to-install MicroSine (a grid tie inverter) fitted on the back of a solar PV panel, which can match the type of electricity flowing though the electric utility grid. Batteries are thus not needed, but the PV electricity ceases if the grid breaks down for any reason.

ENERGY CONSERVATION THROUGH LIGHTING

Utilizing solar energy is always an exercise in energy conservation. Commercially produced energy is expensive, and conservation measures expand the available quantity of electricity and reduce the need for constructing new generating capacity. Solar-generated electricity should not be used to operate appliances, such as hot plates or hair dryers, that use resistance elements, because such appliances draw more electricity than most modest-sized solar systems can make available at a given time. Few people realize that an office copier or clothes dryer uses resistance heating and is an energy-costly device. At ASPI, the PV system is in place and efficiency is the order of the day. We changed all of our overhead and other small lighting units to fluorescent lights, which required abandoning some dimmer systems, giving us an equal amount of luminescence at about one-fifth of the energy demand. And the resulting lighting was softer and easier on the eyes.

Compact Fluorescents

Even in the absence of a PV system, compact fluorescents can conserve electricity. When we hold our hand up to an incandescent bulb, or place such bulbs in baby chicks' brooding houses, or use lights to prevent pipes from freezing, we realize that light and resistance heating occur simultaneously in old-fashioned lighting devices. Unfortunately, traditional incandescent bulbs are cheap, so the general public has not moved rapidly in the direction of commercially available lightweight compact fluorescent lights (CFLs). It is estimated that the conversion of the nation's interior light fixtures to such efficient lighting, in everything from corridor lighting and exit signs to reading lamps and kitchen lighting, would halt the need for any new power plants in the foreseeable future (a savings of forty power plants).

The EPA estimates that if every family replaced its five most-used light fixtures with Energy Star–qualified lighting, a savings of $60 per year would accrue to the average household; taking this step would also reduce greenhouse gas emissions by 1 trillion pounds. CFLs are four times more efficient, use 50–80 percent less energy, and last up to ten times longer than incandescents. Though initially more costly than incandescent bulbs, the CFLs will save money in the long run. The new fluorescents attach directly to light sockets, do not require ballast systems, last much longer, are more sturdily constructed and break less easily than older models of compacts, and are decreasing in price because of increased demand. If an individual is going to undertake only one energy-conserving step, that step should be to replace all lighting with more energy-efficient systems.

Other Lighting Options

This subfield is as complex as PV generating equipment, mainly because changes are occurring rapidly. Several outlets now offer a host of energy-efficient lighting accessories, especially 12-volt DC options; indoor fluorescent lighting of all sizes and shapes; variable-speed switches; Christmas tree and other interior decorative lights; and bullet or halogen lights for bedroom, workshop, or kitchen. The available list of energy-efficient items goes beyond the interior and includes a variety of sensor lights, streetlights, and outdoor path markers with solar-charged batteries. With time, these options expand, prices drop, and lightbulb lifetimes are extended.

PV APPLIANCES

All traditional electric appliances could have PV counterparts:

Solar fans. Perhaps the best-known PV appliance outside of lighting fixtures is the solar fan. Solar fans come in various sizes and are designed for specific purposes. A passive-solar-cooled home may need such fans, especially to fill the structure with cool early morning air in the summer. The compost toilet can be equipped with a PV unit that has no storage but runs only during the daytime, which is sufficient for a unit that is moderately used.

Solar water pumps. The use of PVs is particularly beneficial in parts of the world where irrigation water is of critical importance for producing food and accessing uncontaminated drinking water. The units are a combination of traditional pumps with PV panels and battery-storage systems. If water pumping does not have to be continuous, as in watering livestock or occasionally irrigating fields, costly battery storage systems can be omitted and PVs can directly run the pumps during daylight hours. The PV sources can be mobile units, such as the one shown in figure 1.2, a product of a 2005 Long Branch PV workshop.

Solar water fountain pumps. Interior and exterior water fountains and waterfalls can be more than merely ornamental. They can have therapeutic effects on shut-ins, senior citizens, and hospital patients. Joseph Campbell mentions the sound of flowing water as being one of the primitive sounds that reminds us of our connection to all life forms in the living environment. If exterior displays are operated only in daylight hours, they require no costly battery storage systems.

Solar signs. Illuminated welcome signs that run on solar power are an excellent way to promote solar use. In 2003, ASPI designed a solar sign for the city of Mount Vernon, Kentucky. For some all-night signs, storage batteries are needed; for limited use, less storage is needed and the signs can be turned on or off by light-sensitive timers.

Solar food dryers. Solar food drying is an economic and efficient way to preserve food (see chapter 7). Commercial food

Fig. 1.2. Portable solar generators such as this one can be used to power solar water pumps, power tools, and appliances.

dryers that run on electricity have been available for years, and PV-powered fans for such systems can be obtained.

Solar refrigerators. Solar-powered cooling equipment is of critical importance in remote hospitals and clinics where medicines must be kept refrigerated. High-efficiency commercial cooling units are costly, but the urgency of the need justifies the investment. Small units with limited capacities are now available for several hundred dollars.

Solar communication and computer equipment. Laptop adapters and other solar-powered computer systems are commercially available. Sophisticated signals for communication and transportation (highways, ships, railroads, and airplanes) are now commonly operated using PV energy sources and have proved very dependable.

Electric fencing. One use of solar energy often omitted in surveys is that of electric fencing to keep livestock from wandering. Often fences are erected on terrain inaccessible to public utilities and depend on PV-generated electricity to be effective.

Entertainment and portable equipment. Solar is used in an endless variety of electronic devices. Some of these solar-powered devices are commercially available. A number of solar catalogs (see Resources) reflect the rapid expansion of this field. Solar enthusiasts are encouraged to attend regional solar energy fairs to acquaint themselves with the latest innovations.

NET METERING

Many states have systems in place allowing electricity generated from solar, wind, microhydropower, or other forms of renewable energy to be fed back into the grid either at either wholesale or retail electricity prices. This is especially advantageous for the independent power generator, since it does not demand an array of costly lead-acid or other deep-cycle storage batteries, which are normally required to store surplus solar energy as chemical energy for use when the sun is not shining. In net metering, surplus energy produced by residential PV systems (or other solar sources) is fed back into the electric utility grid to be used throughout the region—and the individual producer's electricity meter runs backward, so that the customer receives credit for the retail value of the donated electricity. While viewing the older utility meter with its circulating wheel under glass cover on sunny days, ASPI staff and visitors could see the wheel tottering as it moved from running forward to running backward. When a standby copier or computers were turned off, the amount of savings could be seen. (The relationship between appliance use and conservation can be learned almost as graphically from interactive computer systems.)

The ASPI solar automobile charging station has been integrated into the office energy supply and into the utility grid. It was a temptation to respond to utility monopolies and their polluting nonrenewable power sources by being a "solar guerrilla," secretly introducing solar-generated electricity into the system without telling the utility, simply letting the local meter run backward with no notice. But utility linemen must be informed about alternative systems, since the small producer is actually generating electricity during normal daylight working hours and linemen could be electrocuted if they thought all power sources were turned off. And in any event, yielding to the temptation is not a way to change the practices of a central supplier of electricity. It is far better to do something about the lack of net metering, enlisting companies that can see promotional, economic, and other advantages and making the public aware through media coverage of renewable energy's potential to supply clean energy to the community.

Utilities get promotional benefit from their sensitivity to small energy producers, but the ultimate benefit to utilities is that during hot summers when loads peak due to increased use of air-conditioning units, individual solar producers are most able to feed back electricity. Generally, states that produce fossil fuels are more reluctant than nonproducing states to encourage federally directed net metering procedures through their respective power commissions. However, ASPI's Joshua Bills ushered in the first such net metering system in Kentucky with the cooperation of the large Kentucky Utilities Company. Other individuals and groups are now joining, and gradually other utilities will be encouraged to start the long journey from generator of electricity to community service provider. Photovoltaics and other renewable energy systems will help change the nature of energy distribution in the coming decades.

CHAPTER 2

Microhydropower

My involvement with microhydropower began in the 1970s while [I was] in college at Georgia Tech. The energy crisis of the seventies got me started looking at alternative energy, and I experimented with natural gas as a motor fuel, which I hoped to replace with landfill-produced methane. I often visited the mountains of northern Georgia and western North Carolina and began to read all I could about microhydroelectric power generation, especially after seeing several high-head systems in use by the utilities in Jackson and Macon counties in North Carolina.

There was major interest in the eighties. After setting up a 10 kW [kilowatt] system at our house on Cherry Gap Branch in Cullowhee, Jackson County, North Carolina, I did a couple of dozen site surveys for others. As a result of the more feasible studies my company, Mountain Water Power Systems, installed about a half dozen smaller systems for others. Interest has been off and on since then. Cost is a concern—a system costs from $4,000 to $20,000 to install, and land with water rights is hard to obtain. Efficiency has been an issue on some of the smaller sites. It is hard to obtain the published numbers since you do not usually have the engineering budget to design and build one-of-a-kind systems so off-the-shelf components must be used. Individuals are sometimes unwilling to make the lifestyle changes that alternative energy requires. Even with 10 kW, my family found it hard to make adjustments when there were three small children and both of us worked outside the *home. You want to turn on the stove, start a load of clothes, and take a shower when you get home!*

There is a great deal of interest lately with the cost of energy, but the public is unwilling to spend a lot of money in return for a small amount of energy compared to what they are used to. Maintenance is a concern for some: cleaning leaves from the trash rack in the fall, for example. Erosion of the runner due to sand from flooding is a maintenance item I have to deal with, but to me it is worth it.

With the cost of gasoline most of the alternative energy interest now is in transportation, but there might be a fit with electric cars. Maybe a hybrid modified for plug-in use might be charged by a hydroelectric plant. Recent developments in electronics and batteries might be leading to a more efficient, trouble-free generation in the future.

Richard Hotaling, engineer,
Mountain Water Power Systems, North Carolina

From earliest times, diverting water for irrigation allowed indigenous peoples to expand food growing into drier areas. Flowing water has been tapped for power for grinding grain and later for mechanical uses such as sawmills. As Norman Brown notes in his book *Renewable Energy Resources and Rural Applications in the Developing World,* the history of small-scale hydropower development can provide many useful ideas on how to aid rural areas of developing countries with water power potential to better provide for their basic needs. In the beginning, waterwheels were able to convert the flow of water into mechanical energy; later the small turbine increased the amount of power that could be generated at a given site.

In the middle of the nineteenth century in the rural areas of the United States blacksmiths and foundrymen began to produce water turbines based on the original French models, creating extremely profitable businesses. Many of the newly designed turbines became known by the names of the American innovators, such as Francis, Kaplan, and Pelton. The mills that were powered by these water turbines showed the potential for rural industry based on decentralized power sources. Brown notes that they "turned out such household products as cutlery and edge tools, brooms and brushes . . . furniture, paper . . . pencil lead . . . needles and pins . . . watches and clocks, and even washing machines." For the farm they turned out "fertilizers, gunpowder, axles, agricultural implements, barrels, ax handles, wheels, carriages. There were woolen, cotton, flax and linen mills; . . . tannery, boot and shoe mills . . . and mills turning out surgical appliances and scientific instruments."

In a bioregion that is blessed with an average of 47 inches of precipitation per year (over 80 inches a year on the western slopes of the southern Blue Ridge) and rugged mountain topography, developing local waterpower resources makes good ecological and economic sense.

MICROHYDROPOWER POTENTIAL

In research done at Appalachian State University and Western Carolina University in the early 1980s, scientists looking at U.S. Geological Survey stream-flow data and computing elevation differentials determined that the mountain counties in western North Carolina could become net exporters of electricity by simply tapping the potential microhydropower sites that could generate between 5 kW and 100 kW of electricity. This estimate did not include systems that could generate more than 100 kW, which would be considered small hydro. Nor did it include sites such as the 3.5 kW site developed at the Long Branch Environmental Education Center in the Newfound Mountain Range at 3,200 feet elevation. Numerous other productive sites of less than 5 kW can be found across the mountains—a virtually untapped renewable resource that could help create the energy needed to develop a robust and vibrant bioregional economy.

SYSTEM DESCRIPTIONS

Potential microhydro sites must be evaluated to determine if sufficient hydraulic head—fall or drop (difference in elevation from the point of intake into the pipe [penstock] to the powerhouse with the turbine / generator)—is available. The quantity of available water, known as flow, is the other critical determinant in deciding a site's feasibility. If a site has sufficient flow, measured in cubic feet per second or gallons per second, along with enough head, a project can be pursued. A typical system is illustrated in figure 2.1.

With the steep elevations in the mountains, high-head, low-flow microhydro systems utilizing impulse turbines such as Pelton wheels have become very popular. These systems are often referred to as "run of river" systems because they do not require large impoundments of water behind dams, and so do not present any of the environmental problems associated with small or large hydro systems. Only a partial and temporary redirection of the stream's flow is required to generate energy, with virtually no environmental impact on a stream.

A small (2- to 4-foot-high) diversion weir can be constructed in the streambed, high on the watershed, to divert a portion of the stream's flow through a trash rack (to keep fallen leaves and other debris out of the system) and then into a settling basin where abrasive soil particles can be settled out. From there the water flows into the penstock that will channel the water under pressure to the turbine in the powerhouse, which ideally is located near the point of end use or near a utility grid intertie. As the water flows, the turbine turns a generator to produce electricity.

All of the headworks need to be carefully designed and constructed with highly durable materials, such as steel-reinforced concrete, to withstand hundred-year or greater frequency rains and flood conditions. The forebay (settling basin and trash rack) needs to be constructed off to the side of the stream's main

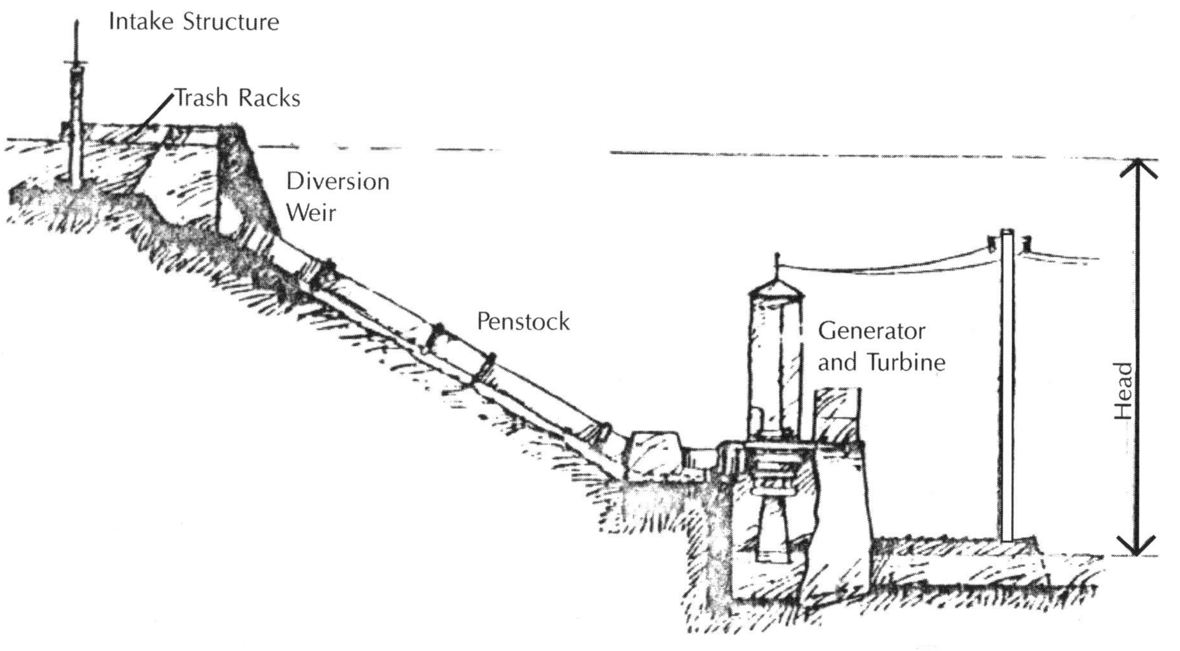

Fig. 2.1. Typical microhydroelectric system

channel so that it is not washed out during intense storms and floods. The settling basin allows abrasive soil particles, grit, or sand to settle out so that they do not abrade the metal of the runner or turbine. The penstock can be made from steel or smooth-walled polyethylene or PVC pipe to reduce friction losses that would otherwise impair the overall efficiency of the system.

CIVIL WORKS

Although, compared to larger systems, microhydro systems do not typically require extensive civil works, the work that is required must be done with careful attention to detail. The major tasks may include building the diversion weir or dam, building the forebay, installing the penstock, and building the powerhouse.

Building the Diversion Weir or Dam

Small diversion dams can be built out of locally available materials—logs, stones, or steel-reinforced concrete. The purpose of the small diversion dam is to pool up enough water so that it may be channeled into the forebay or settling basin before entering the mouth of the penstock pipe. These diversions do not normally have to exceed a height of 3 or 4 feet. Often these small diversion pools can be multifunctional and serve as sources for irrigation, fish farming, watering of livestock, fire protection, or even a cool and refreshing swimming hole. In the construction of the diversion, careful attention must be given to making sure that water will not undermine or circumvent the dam. Also provisions should be made for a spillway across the top and on the downstream side of the dam so that these areas will not be subject to the eroding action of the flowing water. Effective spillway designs can include local rock and concrete.

Building the Forebay

The forebay's settling basin is an important feature of the intake structure, especially if the stream is at all susceptible to frequent or periodic siltation due to soil disturbances or erosion above the location of the diversion. Even cattle grazing on pastureland above an intake structure can create erosion, and sediment can wash into the creek. Grit can cause unwanted abrasion on the penstock, nozzle, and turbine buckets and reduce their life expectancy. The trash rack will prevent larger debris such as rocks, tree branches, and leaves from approaching the penstock and should be designed so that it can be as self-cleaning as possible. A steel rake can sometimes come in handy for removing stubborn trash. In larger streams, it is probably a good idea to install a log boom just above and to the stream side of the trash rack and total forebay. This log or series of linked logs is fastened either to a mooring in earth or to a tree and floats parallel to the stream. It is able to deflect any large branches or tree sections, which might do considerable damage to the intake structure.

Installing the Penstock

Polyethylene or PVC pipe is generally recommended for use as a penstock because of its smooth bore and general availability, although if salvaged steel pipe is available, it is also a durable choice. Where the penstock exits the intake structure, care should be taken to ensure that the pipe is directed out of the streambed proper as soon as possible to protect the pipe from damage by boulders or any other heavy debris carried by the stream in high flows.

To protect the pipe against collapsing in upon itself if the intake is blocked with trash, a small-diameter vent pipe should be installed off the first section of pipe to leave the headworks so that air can be drawn into the penstock to keep it from imploding. The penstock can be run on top of the ground or buried; the buried pipe would be much less susceptible to damage from falling trees, vandalism, or subfreezing weather. Angle bends of 45 degrees or greater in the penstock should be avoided because of friction losses and the undue stress they put on the pipe. Dips in the run of the penstock should likewise be avoided because of the possibility of silt accumulation and the danger that the pipes will freeze if the penstock is drained for winter repairs.

Building the Powerhouse

The powerhouse requires a solid foundation to support the turbine and generator set, especially with the amount of thrust that is applied to the turbine through the nozzle from the high-velocity jet of water. Sufficient room in the powerhouse should also be designated for the installation of electrical equipment.

ADVANTAGES AND CONCERNS

Working examples of microhydropower systems in the mountains of western North Carolina include a 10 kW system on Cherry Gap Branch in Jackson County, a 17.5 kW system on Laurel Creek in Watauga County, a 3.5 kW system on the Long Branch of the Big Sandy Mush Creek in northwestern Buncombe County (see figure 2.2), and a 2.1 kW system on Gashes Creek in Transylvania County.

At less than $0.025 per kilowatt-hour (kWh), microhydroelectric generation is the cheapest source of electricity currently available for Appalachia. Although the initial capital investment is somewhat high, the equipment is highly durable and does not require frequent installation of replacement parts.

As with any renewable energy system, the owner becomes more and more mindful of the overall amount of electricity that can be generated and consumed, and the most energy-efficient

Fig. 2.2. Pelton wheel turbine and generator set for the microhydropower system at Long Branch

appliances and lighting become essential, as does practicing good conservation.

Concerns with fish movement on larger-scale microhydro sites can be addressed with the installation of fish ladders, but microhydro development is usually done on streams that typically do not experience fish migration. As has been demonstrated in the Cascades and the Sierra Nevada, the design and development of components for microhydro systems can become a robust, decentralized industry that could provide highly skilled employment for many of the Appalachian youth who currently seek meaningful engineering or manufacturing careers away from the mountains.

Perhaps a way of stanching the hemorrhaging of our job-seeking, environmentally minded youth to other bioregions would be to launch a campaign, Microhydro in the Mountains: Going with the Flow! Perhaps we can learn to celebrate our natural resources and culture of stewardship. And perhaps we can create a truly meaningful, ecologically conscious, and sustainable economy by combining our technological ingenuity with our deep instincts to care for the Earth.

CHAPTER 3

Wind Power

The gentle shushing sound of our small wind generator is soft and relaxing—almost meditative. We like to think of it as the sound of coal not being mined. In a state ravaged by mountain-top-removal mining, wind power is a practical, ethical, and Earth-friendly choice. Our area of West Virginia is visited every winter by substantial snow and ice storms, during which our neighbors may lose electric power for days at a time. Our alternative power system keeps right on going! We are delighted to be able to express our ecological values in the energy we use.

Carol Warren and Todd Garland,
Webster Springs, West Virginia

Wind power is far from new. In the seventeenth century, England had ten thousand windmills of 10 to 20 horsepower each; in the same period, twelve thousand wind machines were operating in the Netherlands, primarily to reclaim inundated cropland.[1] In the nineteenth century, American wind devices were mostly found on the prairie and on individual farms, where they pumped water for livestock. Many of these devices survived even after cheap rural electrification programs, and some workable relics remain today. Even though the technology of these early devices and those of ancient Dutch windmills was notoriously ineffi-

---

1. Christopher Flavin, *The Turning Point,* Worldwatch Paper 45 (Washington, D.C.: Worldwatch Institute, 1981).

cient, the devices were reliable and satisfactory for the work that had to be done.

Today, wind is the fastest-growing energy source in the world. Almost 11,531 megawatts (MW) of generating capacity was added in 2005, an increase of 40.5 percent over the previous year's increase, with large percentage increases in North America and Asia. Denmark, a longtime leader in wind use (3,122 MW total capacity), now obtains a greater portion of its energy from wind than the United States does from nuclear facilities.

Germany, after taking a number of measures related to renewable energy, including imposing energy taxes on nonrenewable energy sources, has the largest installed wind power capacity in the world: 18,428 MW. Spain (10,027 MW total wind capacity in 2005) has set an updated goal of 20,000 MW in 2011, or 15 percent of its national electricity consumption. In the Netherlands, consumers want green power, and their expanding wind farms cannot meet demand. The European Union uses more than four times the wind power that we do in the United States and reached its 2010 goal of 40,000 MW five years ahead of time. Wind now accounts for 3 percent of total energy use on the Continent.

Europeans recognize that wind has two advantages: it is virtually environmentally benign, and with improvements in technology, wind-generated electricity is going down in price. Unlike nonrenewable energy sources, wind power generation entails no chemical emissions or major land disturbance. And wind equipment today is far more efficient than its forebears of just a few years ago and has advanced from state-of-the-art devices with limits of 550 kW each in 1996 to over three times that limit as of this writing. Wholesale wind-generated electric rates are dropping rapidly; they are now as low as 3 cents a kilowatt hour and will be falling toward the 2-cent range in the coming years. Wind is now competitive with coal and other fossil fuels, and the utility companies seem to recognize this better than our national policymakers.

In the future wind power will be used to break down water and generate hydrogen for the anticipated hydrogen economy. In this regard wind is recognized worldwide as the fuel of the future.

U.S. WIND POWER

> *Without political support, wind energy remains at a competitive disadvantage due to distortions in the world's electricity markets created by decades of massive financial, political and structural support to conventional technologies.*
>
> Arthouros Zervos, president,
> European Renewable Energy Council

Unlike the European nations that committed themselves to reaching Kyoto treaty goals, the United States is playing catch-up with renewable energy applications. Americans are only starting to realize wind's potential (9,149 MW total installed capacity in 2005). Denmark and other European nations are relatively small, compact nations whose population centers are close to wind energy sources; the United States has vast, untapped wind potential in many regions and near some but not all population centers. Transmission costs can be high to reach all population centers.

That renewed interest in American wind power has gone big time is demonstrated by the large wind farms at Trent Mesa and Indian Mesa, Texas, and Altamont Pass (where siting problems caused mortality in the golden eagle population) and Palm Springs, California. These are not backyard operations by small farmers as in times past, but the work of major utility and energy companies, such as American Electric Power, which are now prepared to enter the wind-generating picture. Appalachians and others are asking whether the wind must be tapped only by centralized electric utilities, or whether wind power, like decentralized solar energy systems, can serve individual households as well.

Wind power holds great promise in many parts of our nation, though not always where the heaviest concentrations of people are located. The best wind power sites, classes 6 and 7, are found throughout the upper portion of the lower forty-eight states and in Alaska, along the northern Atlantic and Pacific coasts, and the Gulf coast of Texas. However, the major concentration is in the sparsely populated Great Plains.

While the greatest wind power potential is in less-populated places, a surprising number of population centers are near wind potential areas—San Francisco, Milwaukee, Omaha, Kansas City, Oklahoma City, Dallas–Fort Worth, Minneapolis–St. Paul, and San Jose. North and South Dakota, Nebraska, Kansas, Oklahoma, Texas, Iowa, Minnesota, Missouri, Wyoming, Montana, and much of Wisconsin and Colorado (54 million people) could meet all their energy needs with safe, nonpolluting wind energy without any major difficulty. The American Wind Association estimates that by 2020 some 5 to 10 percent of electric needs could be met by wind power, if a concerted effort is made using current technology.

But why set targets too low? Some experts estimate that the wind potential of just the states of North and South Dakota and Texas would be sufficient to power the current extravagant needs of the United States (not taking into account transportation costs).[2] The potential for wind power is so huge that coal-, oil- and natural gas–fired plants could be retired.

APPALACHIA AND WIND POWER

Only a decade back, even renewable energy advocates in Appalachia were not fully aware of the region's wind power potential. However, people are now taking note that certain

2. D. L. Elliott et al., *An Assessment of the Available Windy Land Area and Wind Energy Potential in the Contiguous United States* (Richland, Wash.: Pacific Northwest Laboratory, 1991).

areas of the Rocky Mountains and the Appalachian Mountains also have class 6 and 7 sites, especially in western North Carolina and parts of central and eastern West Virginia. Wind enthusiasts find that many of the higher elevations consistently have high-velocity winds and thus are well suited for electricity generation. Wind experts say that 2 percent of the land area in the twenty-four western North Carolina counties (138,000 acres of ridgetop land) is suitable for utility-scale wind projects; this potential area is halved if one excludes federal and state forests and parklands, viewshed buffers, and the Appalachian Trail.

Advances in technology and design, such as longer blades and better gearing mechanisms, now permit the use of lower-velocity wind.[3] So areas in the central Appalachian range that are in wind classes 4 and 5 are now suitable for wind power generation.

Because wind classes can vary greatly with slight differences in elevation on a mountainside, data must be collected to determine the exact location where a unit should be placed in steep mountainous terrain.

Small- versus Large-Scale Applications

Appropriate technologists generally follow the philosophy of E. F. Schumacher and think that smaller is better. With wind power, one could speak of large-scale electric generation and focus on flat plains or off-coast wind farms where suitable territory is plentiful. Larger-scale Appalachian wind farms are feasible, as is evidenced by the Tennessee Valley Authority wind farm on Buffalo Mountain in central Tennessee (figure 3.1 shows a TVA wind turbine in Tennessee). Others are being planned or developed in east-central West Virginia. Some people regard this as a detriment to the scenic beauty of the landscape, but they also doubly regret each new fossil fossil-fueled power plant with

3. Christopher Flavin, "Wind Power Rise," in *Vital Signs 1994*, ed. Lester R. Brown et al. (Washington, D.C.: Worldwatch Institute, 1994).

its added deterioration of Appalachian air quality. We must still emphasize small-scale wind facilities, whether of the traditional free-standing kind or the tilt-up varieties that are fastened with guy wires. Small-scale wind power is coming to the region.

Older wind enthusiasts speak respectfully about Marcellus Jacobs, the father of American wind-generated electricity. The small-scale machines he designed and built during a quarter of a century of manufacturing acquired an excellent reputation for durability, and some still operate on farms. They were generally placed near homes to minimize transmission-line losses and were hooked up to lead-acid batteries for charging and storing energy to provide electricity. Today many of these less-efficient wind generators rust at old farmsteads and are being replaced by state-

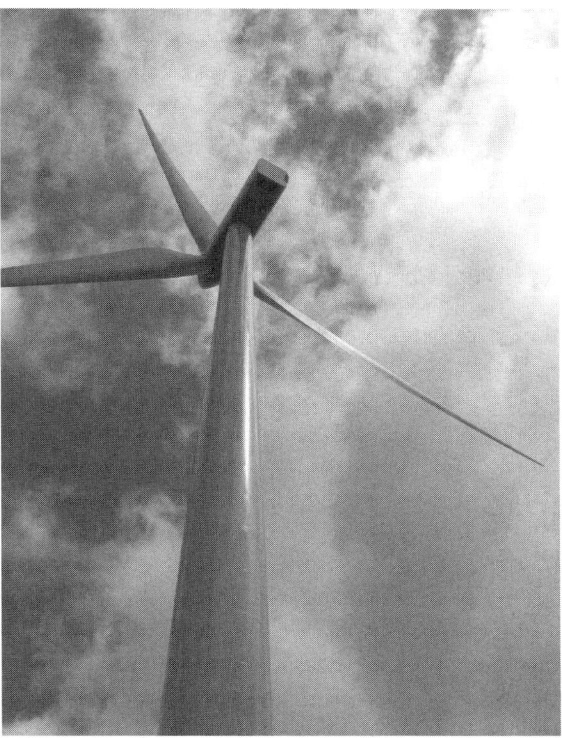

Fig. 3.1. A TVA wind turbine in Tennessee

of-the-art, aerodynamically engineered devices that can run at far lower wind speeds (as low as 5 to 10 miles per hour). These newer varieties are constructed at ever-lower costs for use by both large and small systems. If the energy playing field were leveled through governmental incentives and tax breaks, wind would be quite competitive with nonrenewable sources of electricity at both the utility and the domestic scale.

Professor Dennis Scanlin at Appalachian State University at Boone, North Carolina, a solar and appropriate technology expert, is enthusiastic about wind power potential in Appalachia. Through funding from the U.S. Department of Energy, he is establishing testing towers in various choice locations in North Carolina and neighboring states. He says that central Appalachia has far greater wind power potential and promise than had previously been thought. He has already established a number of anemometer/data-collecting towers and is gathering data at higher elevations over several-year-long periods. The results do not have to be directed only to commercial wind-generating facilities but may also apply equally to residential and small-business wind turbines, which are becoming more feasible over time.

Appalachian Wind Concerns

Appalachian wind sites may be on public land or on private land whose owners are wary of renewable energy development. Tower costs can escalate if the testing unit or wind generator tower must be brought in over difficult terrain. Accessibility to the site is a major concern. Where land is under a conservation easement, variances may be obtained by persuading the trustees that wind power reduces fossil-fuel use and thus ultimately enhances the conservation value of the flora and especially the forests. Some of the best North Carolina wind sites (class 7) are on portions of the Long Branch property under conservation easements, and a variance for renewable energy development is necessary. The following are some barriers to development of wind power:

Aesthetic objections. A major barrier to wind development is that residents in areas of high wind potential do not want the landscape (mountains or coastlines) marred by "unsightly" wind turbines. These misinformed property holders wish to see what from their vantage point is an undisturbed landscape, no matter how much pollution has resulted from fossil-fueled plants and how degraded the forest landscape may be by acid deposition and ground-level pollution. They may even try to masquerade as environmentalists, but their ultimate agenda is to preserve their own private domains while allowing the public commons to deteriorate. What about people affected elsewhere by coal strip-mining operations, fossil-fuel pollution, or high-level nuclear wastes? An environmental consciousness must reach beyond trying to preserve a private viewscape of a terminally ill forest ecosystem. Besides, wind farms can be attractive. Wind generators can be sited to minimize forest fragmentation caused by the construction of transmission lines and access roads in forested areas.

Avian mortality. Closely related to threats to scenic landscape is the problem of bird kills at wind turbine sites. The early unfortunate siting of a wind farm at Altamont Pass, California, was done without ecological due diligence. The area, unbeknownst to wind developers at the time, was a prime nesting area for golden eagles. To its credit, the wind industry has been honest about its mistakes and is now doing extensive ecological research on all newly proposed sites. Raptors sometimes perch on wind towers and look for prey below. After sighting its prey, the eagle or hawk may make a swoop for the prey and be killed by the blades of the wind turbine. These dreadful accidents number in the dozens of kills each year, while collisions with windows, buildings, communication towers, vehicles, and other man-made artifacts run into the millions each year. High-priority research on noisemaking devices and construction techniques to discourage bird roosting and collisions is under way. Ideas on how to eliminate bat mortality at wind sites is also a top research priority for wind developers.

King Coal. Central and northern Appalachia are coal-producing regions. Promoting renewable energy alternatives, whether solar or wind, in these areas is somewhat difficult. Net metering, which applies to wind as well as to solar energy, is not as popular in coal country as in energy-short regions of America, though several companies are now moving to net metering in Kentucky, North Carolina, and other states. Much still has to be done to modify state regulations and to encourage regional utilities to cooperate in promoting wind energy.

Mountaintop views. The state of North Carolina does not allow construction on mountaintops. Those who want scenic views would not object to such regulations with respect to the construction of motels and houses that mar the landscape. Yet, in France and other parts of Western Europe many of the hilltops have ancient castles or ruins going back a thousand years. It is not inherently wrong to use a mountaintop, if it's for a good cause, such as reducing acid rain and the destruction of the surrounding forests. The small stone tourist tower on top of Mount Mitchell, Appalachia's highest peak (6,684 feet), is far less distracting than the thousands of dead Fraser fir trees surrounding the peak and mountain range. The red spruce / Fraser fir ecosystem is the second-most-endangered ecosystem on the entire continent of North America. The exact causes are considered to be a combination of insects, funguses, tropospheric ozone, acid deposition, and possible intensified ultraviolet radiation caused by thinning of the stratospheric ozone layer. Surely wind-power generators, which would reduce the burden of air pollution that weakens the immune systems of the trees, would help improve the conditions for a thriving forest.

RESIDENTIAL WIND POWER DECISIONS

Throughout the country, individual property owners are increasingly interested in wind power. What about investing in one's own property in true appropriate technology style and not waiting for the commercial utilities to take the lead? Here are some factors to consider in determining whether a particular residence is suitable for wind generation:

Siting. The exact location of the wind tower turbine is quite critical. Some initial testing must be done to determine, with the advice of suitable experts, whether a particular property is suitable for wind generation and at what precise location. Owners can contact the state energy office to obtain a mobile anemometer to use in site selection. A consensus is now emerging that the tower should be close to the point of end use to minimize transmission losses.

Accessibility. Being able to approach the site by vehicle is advantageous for both construction and ongoing maintenance. Using a helicopter to bring in components would be rather expensive, and backpacking them to the final site may be somewhat challenging.

Choice of equipment. Commercial wind power products are multiplying quite rapidly, and final selection will take at least as much thought as choosing an auto or house, or more. Purchasers should consult a trusted wind power expert to help select the correct system. Complete residential wind-generating systems range from $1,300 to $10,000 (2006 estimates). Just as in making residential solar choices, purchasers should determine an approximate size for the wind device that allows for suitable energy conservation while meeting basic housing needs (using energy-efficient appliances, compact fluorescent lightbulbs, energy-saving computers, and ideally very little resistance heating).

Solar/wind combinations. Wind is one component of a total renewable energy picture. Solar energy reaches its prime during sunny weather and on long summer days; wind availability is somewhat more variable. The two can complement each other as part of a single wind/solar hybrid renewable energy system.

Grid connections. Net metering with the existing utility grid could save the cost of banks of batteries such as the less-efficient but highly recyclable lead batteries. If net metering is not presently available in a given area, as much citizen effort should be devoted to implementing this policy as to installing single wind power units.

Lightning protection. Appalachia is prone to violent electrical storms, and structures are generally not sited on prominent ridges and peaks. Without proper electrical engineering, a wind tower situated on the top of a mountain could be an ideal lightning attractant. Not only must towers be securely grounded, but careful consideration must also be given to ensure the safety of any buildings in their vicinity. A lightning strike is powerful!

CHAPTER 4

Wood Heating

My wife and I believe that we are fortunate to live in a region where wood is such a readily available resource. Trees harvested from our property allow us to heat our house, and scrap lumber hauled from a local sawmill provides fuel for our pottery's kiln. It isn't always easy or convenient to use wood but we feel that the efforts are justified. The benefits are greater than simply being an economical alternative to fossil fuels, serving as well to form a broader pattern of daily practices, which allow us to live at a pace more attuned to the seasons.

George Rector, organic gardener and homesteader, Cullowhee, North Carolina

Wood, like solar and wind energy, is renewable and may be an Earth-friendly secondary heat source. The wood user feels a certain control over the heat source, especially when public utilities falter and fail. Burning wood gives a sense of coziness and satisfies the basic human instinct to be mesmerized by fire. The sight of wood smoke curling in the winter valley and the smell of wood smoke have been cherished for centuries, but in recent times, as those good sights and smells give way to signs of air pollution, wood burning has lost some of its aura of Earth-friendliness.

TYPES OF WOOD HEATERS

For thousands of years fire has been used in both cooking and heating. Unfortunately, most heat from open fireplaces simply goes up the chimney, and precious wood has been wasted through inefficient heating methods. We discuss below four methods of burning wood, the first two of which are modifications and improvements in design by Appalachian-based appropriate technology groups.

Finnish Contraflow Masonry Heater

The masonry heater has been known in northern Europe for centuries for its energy efficiency and steady moderate heating. The airflow is such that most of the retained heat is transferred to the large thermal storage mass, or masonry. These masonry heaters are aesthetically pleasing in any setting, especially if glass viewing doors are included in the design.

The masonry stove can be owner-built at low cost, from locally available materials, so that money stays in the locality. Construction costs can be reduced by using scrap metal, recycled brick, native stone, or Cinva-Ram earth bricks (see chapter 21). This heater produces virtually no tar, soot, or creosote, thus reducing the risk of chimney fires. An average-sized home can be heated to comfort with a cord of wood per heating season, far less than conventional wood heaters. This heater can also be designed to bake with an oven, to cook with a metal top, and to heat domestic hot water through an integrated loop of pipes inserted in the masonry walls. Lower surface temperatures on the masonry walls make this heater safer when children are nearby.

The "contraflow principle" is the secret to the masonry heater's efficiency. Directly above the firebox is the secondary combustion chamber where gases, which contain up to two-thirds of the potential heat energy, are burned when they are not combusted directly in the firebox. Combustion temperatures of 1,650 to 1,800 degrees Fahrenheit ensure a complete burn without formation of creosote. The Finnish contraflow masonry heater also contains two vertical baffles that ensure that the heat from combustion is absorbed by the masonry wall of the heater

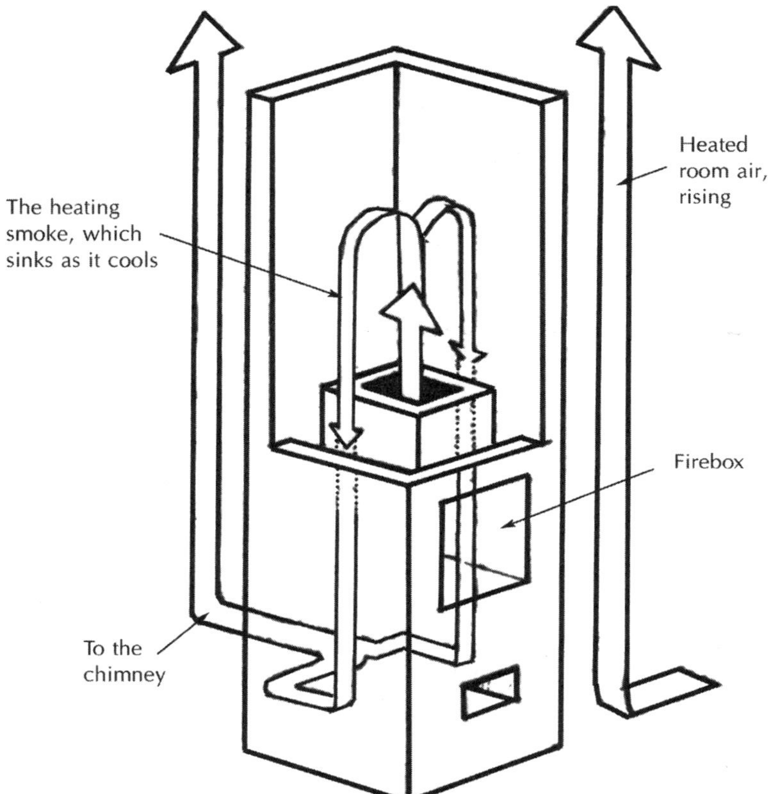

Fig. 4.1. Air currents in a contraflow system. Smoke is pulled downward from the airflow across the top of the chimney.

and is radiated to the living space. One damper is located at the top of the secondary combustion chamber to facilitate drawing smoke when the fire is initially lit; it is closed after the wood has finished burning to prevent heat from escaping up the chimney (see figure 4.1).

The contraflow principle involves a downward flow of heat in the unit from the secondary combustion chamber to the flue located below the level of the firebox. This downward heat flow contrasts with the movement of cooler currents of air on the floor in the living space, which, when they reach the masonry heater, begin an upward convective flow. In this way, hot spots immediately adjacent to the heater are eliminated, and a more even and complete heating of the entire living space is achieved. The Long Branch Community Center is equipped with such a heater and has had satisfactory service for several decades.

The masonry heater ideally should be centrally located to supply heat for an entire dwelling. The location of the foundation and chimney is a deciding factor in the placement of the heater, as is adequate space: the heater has a footprint of approximately 4 feet by 4 feet. Because of the heater's weight, it should be placed on the ground floor, on a foundation of reinforced concrete or bedrock. It is better to place the heater as close to the center of the building as possible to take advantage of convective airflows in the living space, and to avoid placement on outer walls.

Elbow Torch Stove

Mark Schimmoeller, a Kentucky homesteader and periodic ASPI overseas volunteer, designed a very simple, efficient stove for use by inhabitants of Honduras and elsewhere who burn twigs and small wood pieces. This heater concept expands on the rocket stove designed by Larry Winiarski of the Aprovecho Research Center in Oregon. The elbow torch stove uses far less fuel and pollutes less than outdoor grills that are typically used at campsites or in backyards. In developing countries extensive deforestation has made fuel wood scarce. Thus very poor Latin Americans resort to eating only tortillas and not the much-needed protein-rich beans, which take more cooking time. A typical family uses one donkey load of wood per week, but with this elbow torch stove the need goes down to one load every five weeks.

Essentially the elbow torch stove consists of a commercially available metal flue elbow that serves as the firebox. The elbow is embedded directly into a box with wood ash or other noncombustible insulating material. One opening is in the side of the box and the other is the top of the chimney where the box

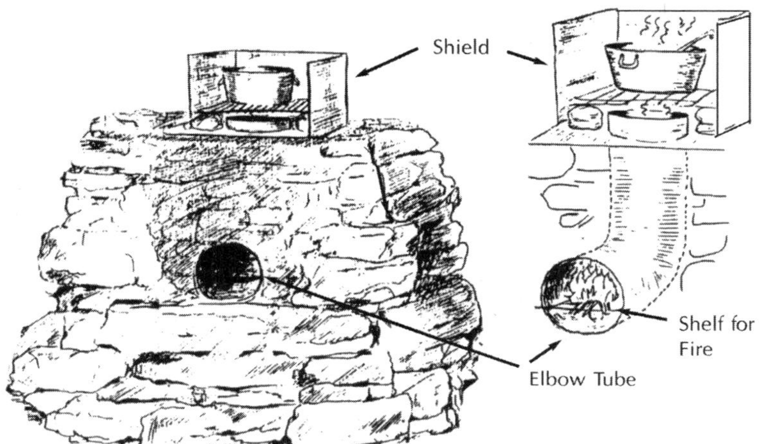

Fig. 4.2. Elbow torch stove

lid would be (see figure 4.2). A fitted metal shelf to hold burning twigs is inserted into the horizontal portion of the elbow with 1-inch clearance beneath airflow. This shelf can be fashioned from scrap metal. Around the top of the vertical portion of the elbow is a fitted metal sheet surrounding the elbow hole and covering the lid portion of the box. The elbow hole is covered by a lightweight grille, which is secured about an inch above the top of the elbow for airspace to keep the cooking vessel from snuffing out the flame. An elbow torch stove fire can cook food amazingly quickly and with very small quantities of twigs. The air shelf in the insulated elbow firebox allows for a natural draft to be created. The low mass of the insulation encourages most of the heat to go directly to the cooking vessel, and the metal cooking top helps directly heat the bottom of the cooking vessel as well as its sides.

The elbow torch stove is extremely low cost (about $5 at most), uses small amounts of wood, creates focused intense heat, takes little space, is portable, can be used instead of outdoor grills in camping areas, saves the forests of the world, and can be built with few technical skills. Directions for construction are found in ASPI's Technical Paper 39, *Elbow Torch Efficient Wood Cooker.*

Woodstoves

Many older woodstoves and ranges do not meet federal government emission standards; they have come to be considered sources of air pollution because their smoke contains carbon monoxide, particulate matter, and tars, which are harmful to human health. Newer, more efficient, and EPA-certified woodstoves, especially those built after July 1, 1992, are far less polluting. Those built to revised government standards may be equipped with catalytic combustors, which burn particulate matter and other gases before they escape up the stovepipe. The catalysts need to be replaced every few years, and they can be easily poisoned by burning nonwood materials. Catalytic devices can be retrofitted into existing woodstoves. Since 2002, the EPA has determined which areas of the country are not in compliance with air pollution standards and has given these areas a limited time (usually three years) to come into compliance. Woodstove standards are promulgated accordingly.

Fireplace Inserts

> *The user of the fireplace comes to understand that fire, like the sun, is a life-sustaining and renewing force, that the real purpose of a fireplace is to renew the energy of those who gather around it.*
>
> A. Barden and H. Hyytiainen

Fireplaces are as old as housing itself, but they can be terribly inefficient. In American homes built at the time of European settlement in the 1700s, all of the fireplaces were net heat losers, taking out more heated air from the interior of the place than they replaced. They generally heated only the person standing right in front of the fire. But the blaze looked cheery, and the radiant heat gave a bit of temporary warmth. Ben Franklin saw this inefficient use of fuel and devised the stove known by his name. From his time on, metal inserts were installed, and modern fireplaces (where allowed by building regulations) are made with built-in air-convection ducts and thus become a modified but far less efficient masonry heater.

Fake-log-filled fireplaces are more decoration than heating device. As one Canadian researcher, A. C. S. "Skip" Hayden, head of Energy Conservation Technology at the CANMET Energy Technology Centre in Ottawa, Canada, notes in his article "Fireplaces: Studies in Contrasts," "Gas firelogs are inefficient, can result in chimney degradation, and can cause severe indoor air quality problems. They should not be installed in today's housing."

GOVERNMENT EFFICIENCY STANDARDS

Due to the respirable particulates in wood smoke and the effects they have on residents and the environment, the EPA has developed certification requirements that govern the manufacture and sale of woodstoves and fireplace inserts for stoves produced after June 30, 1988. Appliances that do not meet these specifications cannot be sold in the United States. The most effective means to reduce air pollution from wood-burning appliances is to buy a new, efficient, EPA-certified woodstove or to install a catalytic combustor on an existing stove. However, these catalytic devices are temperamental and need to be inspected three times each heating season.

Some stoves burn fuel more efficiently than others. Smoldering, heavy-smoke-producing fires may be caused by wet wood, an overloaded chamber, improper lighting, over- or underuse of kindling, or an inadequate draft. The chimney of an inefficient stove gives off dark smoke, though small amounts of dark smoke occur briefly at the start of any fire, but should not continue. Native

American expert fire makers suggest knocking back the chill on autumn and spring days with "grandmother fires," small, hot fires made from a few small pieces of wood and restricted airflow. These experts add that if a house smells of wood smoke, something went wrong in the firing process, the heater is not operating properly, or the flue needs to be cleaned. If the flue is dirty, it is time to buy a new, advanced, low-emissions system or to construct a masonry heater that is truly classified as appropriate technology and can be constructed through a community building project. With a masonry heater, emissions can be reduced by 90 percent and wood consumption cut by as much as a third.

FUEL AND HEATER USE AND MAINTENANCE

Wood-burning devices require some care and maintenance. Too often the public expects the commercial utility person to come and do the necessary maintenance, as with a fuel oil or gas furnace, and that is not the case for most wood appliances; the user is responsible for safe and clean operation.

Efficient operation begins with the selection and preparation of fuel wood. Heaters need good wood to burn efficiently. Fuel gathering and preparation is a late winter or early spring operation for the upcoming heating season (see figure 4.3). Humidity and moisture levels differ widely, as does the time necessary for drying. Moisture content will normally drop from about 50 percent when the wood is freshly cut to 25 percent or less when it is stored in a well-ventilated place. The fuel wood is best cut to stove length, split to less-than-6-inch-diameter chocks, and stacked and stored for future use under a cover that allows airflow, not in a basement or confined space or a tightly fitting plastic tarp. Pressure-treated wood and painted wood (the emissions from which are quite toxic), driftwood, and garbage should never be burned, though dry twigs are good kindling. Wood waste makes good compost, not good fuel wood, except for efficient industrial furnaces at wood-processing locations.

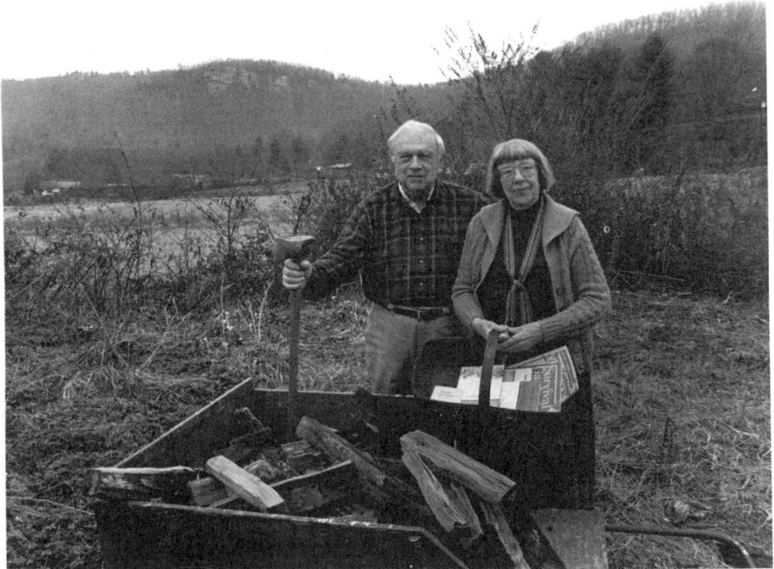

Fig. 4.3. Author John Freeman and the late South Carolina poet laureate Grace Freeman with wood for space heating their home.

Many types of hardwoods, including oak, hickory, wild cherry, black locust, and fruitwood, can be used for fuel. Softwoods, lighter hardwoods, and sap-filled wood do not have the same Btu or heat values and may cause creosote buildup in chimneys. Dried evergreen is good for outdoor fire pits but is not recommended for indoor use. Small branches of fuel-worthy wood can be used in 3-inch or greater widths, but one should split thicker wood chunks. For splitting, a "wood grenade"—a hardened metal, cone-shaped wedge splitter—used together with a heavy maul, "go-devil," or sledgehammer will make this activity an exercise in almost instant gratification.

Many fires are caused by faulty or creosote-clogged chimneys. The chimney determines proper venting of the wood heater, and frequent inspections are absolutely crucial to continued safe operation. Whenever possible, the chimney should be built in the interior of the house to minimize heat loss on outside wall areas and to maximize radiant heat and convective circulation throughout the building. When constructing a building, the contractor or home builder will often have to follow strict building codes and extend the chimney at least 3 feet above the peak of the house. It cannot be stressed enough that the chimney must be well maintained. Flues can be faulty for a number of reasons: cracks or holes, incorrect installation, improper size, or wrong angle of insertion into the chimney.

If a nonpermanent installation is contemplated, the heater should be where residents spend most of their time. Often the wood heater supplies heat for an entire house, and thus a central location is critical for maximum comfort. In essence, the location where a properly functioning chimney can be installed will determine where the heater is placed. Every room with a heater should have a working fire extinguisher within easy reach.

Wood heating takes some careful attention to all phases of operation. Ashes need to be removed, taking special care not to leave hot ashes inside the house to prevent carbon monoxide emissions. When spreading wood ashes outdoors, be sure that the ashes have cooled sufficiently so that no sparks could kindle a forest fire. The wood pile must be kept dry and free of termites; heaters should be checked to see that all parts are functioning properly. The flue and chimney require at least an annual checkup before the heating season to check for loose masonry and ensure that the roof flashing and chimney cap are sound.

LIMITATIONS OF WOOD

In many parts of the eastern United States, forested acreage is increasing, but that does not mean that more people should heat with wood. Forests offer many other ecological advantages that outweigh furnishing fuel wood, although it takes time for the trees to mature. It is possible to find fuel wood without destroying forests. Often trees are cut during development activities and road-building operations, or because they have been damaged by ice and windstorms or are candidates for tree thinning. Shrewd and observant wood burners will likely find trees that property owners will happily give to them as fuel wood rather than pay professional tree trimmers to have them removed.

"Solar first, wood only as a backup" would be a good motto. Wood heating should be considered part of a domestic heating mix and a backup for other sources of heat in colder weather. It is more efficient energy and resource management to consider passive solar heating as the primary heat source and wood burning as strictly secondary.

Underground and earth-bermed houses can also benefit from a small backup wood heater, though elements of passive solar design are often included in these buildings as well. This practical principle of efficient energy and resource management can hopefully be applied to all new home construction.

CHAPTER 5

Solar Heating Applications

When we redesigned and remodeled to incorporate into our new passive solar addition overall improvements in insulation, airtightness, daylighting, appliance and lighting efficiency, we also added as many efficiency upgrades to the older parts of our home as possible, including a new more efficient heat pump. My office, like my household, prefers and uses as much used, refurbished, and high-recycled-content products and materials as possible. Many of our waste materials are recycled or composted. And yes, we do hang-dry our laundry when outdoor conditions permit, since each avoided use of the electric dryer saves 4–5 kWh, associated with over 10 pounds of carbon dioxide emissions when kilowatt hours come from burning coal.

John F. Robbins, certified energy manager,
Morningview, Kentucky

Sunshine is in plentiful supply in Appalachia much of the year. As of now, solar energy is free for all to use. It warms the Earth and delights our souls, allows plants to grow, and gives us light by day. In solar heating applications, the basic methodology is to capture the sun's rays, convert them to heat, and retain that heat as long as possible through heat-storage systems and specific conservation measures.

Solar energy offers several general advantages:

Once the application is built, there are no additional fuel bills, for solar energy is free.

The energy is renewable and comes at no cost to the environment, whereas nonrenewable fossil fuels result in air and water pollution.

There are no worrisome long-lived toxic waste products as in nuclear power generation.

The process offers opportunities for decentralized energy sources and is not dependent on sometimes fallible centralized utility systems.

Solar applications make users continually aware of the need for conservation and prudent use of natural resources.

Additional advantages are listed in the appropriate sections.

SOLAR WATER HEATING

A majority of the world's population does not bathe with domestic heated water but washes with cold or tepid water. Domestic hot water, regarded as a necessity in the developed world, is considered a luxury by many. Water heating accounts for an estimated one-tenth of domestic energy consumption (some estimate upward to one-quarter) depending on use patterns. The developed world uses enormous amounts of heated water, often wasting it by needlessly running faucets and showers before and after use. While avoiding pollution and at the same time building economic infrastructure, much domestic water heating could be achieved by solar heating devices.

Certainly, solar hot water systems are a proven long-term application. Photographs of the April 18, 1906, San Francisco earthquake show damaged buildings with solar hot water systems on their roofs. Solar heating of water began long ago when people set containers of water in the sun to warm. Many of the world's people could obtain their domestic hot water through easy-to-build simple devices. In all respects, solar hot water is the premier solar application, the one that experts regard as having the fastest payback and using the least amount of natural resources.

The basic question for a household is how much hot water is needed. In households where not everyone showers at the same time of day, a domestic water system could have far less storage capacity, and cost less, than if much hot water is demanded within a short period. Electric or gas-fired units that heat on demand can be integrated into the domestic water system; these units save energy when compared to gas or electric water heater storage tanks. However, because they use nonrenewable energy, they are not nearly as friendly or efficient as solar water heating systems with storage tanks of any size.

Solar water heaters must be located on or near the portion of the building that faces south or slightly southwest or southeast, and the site should be accessible for initial construction and for later maintenance. The design should be visually pleasing and in harmony with the building's architecture, and the heater should be constructed to withstand typical wind speeds.

To avoid leaks, care must be taken in combining metal and plastic materials for piping. Since water-storage systems are heavy (a 100-gallon system could weigh half a ton), placing large systems on roofs or in lofts could require substantial reinforcement of roofing or flooring systems. For retrofit situations, the system can typically be positioned on the ground next to the building. A conventional heater may serve as a backup where normal winter sunshine is sometimes diminished by storms and overcast weather conditions. Systems can be installed to generate heated water throughout cold weather, but they require sufficient insulation and protection from the wind, perhaps by dense landscape plantings. Systems in which glycol serves as the primary heating liquid (where heat is transferred to water in a secondary loop inside a structure) can be used throughout winter.

The simplest water heating method is to hang solar water bags (1 to 5 gallons) out on hot days; this system allows comfortable showers that are best enjoyed at the end of a good sunny day. In countries with lots of sunshine and plentiful warm weather, dark-colored water-storage tanks on roofs can be used.

A more sophisticated but still simple passive solar water heating design is known as the breadbox, batch, or integrated domestic solar water heating system (see figure 5.1). The Tennessee Valley Authority has published a design for the breadbox water heater, which consists of a single tank or an array of black-painted storage tanks embedded in a sturdy box with insulated sides and bottom and with a well-fitted and sealed, double-glazed transparent glass facing. The box may have a dark-colored insulated lid that can be manually closed at night to keep water hot longer, especially in cooler springs and autumns.

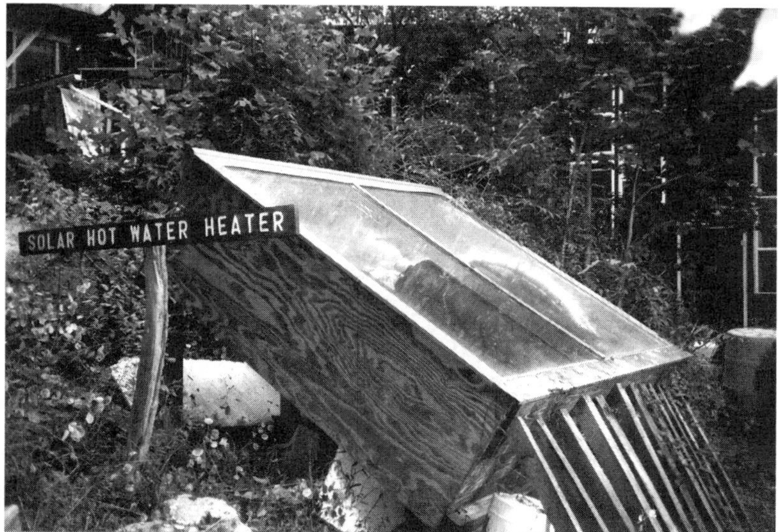

Fig. 5.1. Long Branch's solar hot water heater is of the breadbox type.

This heating process is quite straightforward and is known as a "direct-gain" system. Cold water enters the heat-absorbing tank at the base. As the water heats, the hot water rises in the tank through thermal stratification, and the hot water then flows to the point of end use. This type of system is by far the most economical: the material costs as little as $200, and the system can be built for even less if recycled water tanks are used. A do-it-yourself solar water-heating project requires ordinary building skills and results in an attractive, easily maintained device. Modified designs are available through Volunteers in Technical Assistance (VITA) and the National Center for Appropriate Technology (NCAT) (see Resources).

In contrast to these passive systems, there are also active solar water heating systems. Some solar water heating systems, especially those from the 1977–1986 era, used hard scaling or acidic water in closed-loop systems. These could rupture and burst absorber plates in collectors. These malfunctioning devices were not only high priced but also left local plumbers and domestic users baffled. As poor models were discontinued, better-functioning active or passive systems came on line. In 2003, through a solar workshop, ASPI installed a roof-mounted system that furnishes hot water year-round. This system consists of a glycol primary heating liquid, which passes through copper tubes in a roof array. Its heat is exchanged with water flowing into the building to hot water tanks in a normal plumbing fashion. Such systems do have pumps and pop-off valves, for the solar heated water gets extremely hot in the summer sun. These modern systems have methods to regulate the glycol loop to control water temperature.

As with all appropriate technologies, basic maintenance is needed, such as cleaning the glass cover as frequently as conditions require. Well-insulated breadbox water heaters can sustain temperatures below 25 degrees Fahrenheit for periods of time especially if fitted with a lid that is closed at night. All heating systems should be checked for leaks and, where necessary, glazing areas should be caulked.

SOLAR COOKERS AND OVENS

Solar food cookers could potentially save forestlands, in part because over a billion people depend on trees to furnish the fuel to cook their food. In some countries, people with limited resources spend up to a quarter of their income on kerosene or other fossil fuels for cooking, and sometimes spend one day a week collecting firewood. Since so much food could be cooked for longer periods at lower temperatures (rather than by high-heat frying), the potential for fuel savings from solar cookers is immense. Solar cooking is smokeless and thus prevents the lung damage that often occurs in smoky kitchens. If the cooker is large enough, several dishes can be cooked at the same time. Furthermore, the cooker can be an easy, low-priced way to purify water in areas at risk of intestinal disease.

The solar oven must be placed in a location that receives sunlight for a number of hours, depending on the time of year and geographic zone; solar ovens work well in the tropics year-round. When the sun is one-third of the way from horizon to overhead, many easy-to-cook foods will be ready at midday. The closer the sun is to 90 degrees (directly overhead), the greater the cooking power. The most intense cooking could take place from two hours before to two hours after noon sun time. In temperate zones the angled winter sun is weaker. Dust and smog will slow cooking time, but generally the time is unaffected by wind, humidity, and outdoor temperature.

A solar box cooker or oven is an insulated box within a box that contains a dark cooking pot to absorb sunlight (converting light to heat); shiny surfaces of aluminum or Mylar on reflectors and inner box surfaces to reflect rather than absorb energy; and a tight-fitting, transparent material through which sunlight passes and which traps the longer-wave heat energy. The solar oven is equipped with adjustable reflectors that are directed at the sun. Maximum effect is achieved by readjusting them to obtain direct sunlight throughout the cooking process. However, allowing longer time for cooking will save the cook from having to watch constantly.

Solar cookers may be permanent or temporary. The former are more durable, and the latter are portable and even, in some cases, collapsible, so that they can be folded for easy transport. Solar cookers of whatever construction are easy to maintain, but lighter-weight units should not be left in the rain; all moisture should be wiped away after cooking so that the box and insulation stay dry. Smaller units constructed with cardboard or light box material are not as durable as those made of plywood and wood, especially when frequently used. For frequent community oven use, the units can be made of masonry, brick, pressed earth, or adobe.

A variety of lightweight do-it-yourself units can be constructed from simple cardboard boxes, aluminum or Mylar sheets, plate glass, a darkened cooking utensil, sawdust or crumpled paper insulation, and sealing materials. Sturdier containers can be made of wood or plywood, and fiberglass insulation, larger reflection plates, and larger cooking vessels can be used to minimize unused heated space. Instructions for making some of these lighter-weight cookers are given in ASPI Technical Paper 1, *Solar Box Cooker;* and Technical Paper 67, *Lilongwe Solar Box Cooker* (see Resources). Overseas work by Mark Schimmoeller and other volunteers on three continents has found that the sturdier ovens are more likely than other solar-energy heaters to stand up under continuing use over relatively long time periods.

The amount of food cooked depends on the size of the box. Most of the smaller units described can cook up to 10 to 15 pounds of food on a sunny day, especially through slow-cooking techniques. Add no water when cooking meat, fresh fruits, or vegetables, and no nutrients will be lost in the cooking process. For cooking dried beans and grains, add the amount of water used when cooking with other energy sources. The gentle cooking temperatures of 200 to 235 degrees Fahrenheit are ideal for cooking food without burning it. Even tough meats are tenderized. Furthermore, there is no need to stir cooking foods. Checking the food will necessitate opening the lid, and the resulting heat loss will slow cooking. For better cooking, move the cooker with the sun, especially in winter or on cloudy days. Lilongwe solar heater cooking times are: beans, 3 hours; rice, 1½ hours; vegetables, 1 hour; beef, 2 hours; and fish and chicken, 1½ to 2 hours.

Solar cookers are quite versatile for people with limited fuel resources, campers, hikers, retreat-cabin dwellers, and refugees. Water can be pasteurized by heating to at least 150 degrees Fahrenheit. Foods (especially acidic nonmeat products) can be preserved in heated, sealed containers in the solar box. When other methods are unavailable, the solar cooker can disinfect medical equipment, bandages, and septic wastes. Solar cookers can serve as "wonderboxes," or wrapped insulated containers, after vessels of food have come to a boil over an open fire or by some other method, for longer-term, slow cooking.

"SOLAR" DRYING OF CLOTHES

Project Laundry List (www.laundrylist.org), directed by Alexander Lee of New England, champions the more traditional method of hanging out clothes to dry on a clothesline. Lee also promotes National Hanging Out Day each April 19. Drying clothes outdoors is frowned upon in modern America, considered something only poor folks would do. This traditional practice is falling out of favor in much of Appalachia as well; only rarely does one see clothes on a line, except in certain rural areas of the region. Much has to do with the coupling of electric washer/dryer operations both in American laundromats and in private residences.

In this country, according to the Rocky Mountain Institute, about 6 to 10 percent of residential energy goes toward running clothes dryers. The United Nations Environment Programme estimated in 1991 that the average American uses more energy to dry clothes than the average African uses for all purposes throughout the year. All of this residential energy use could be avoided by taking advantage of the sun and fresh air when weather permits. There would be a massive savings of nonrenewable energy (using renewable energy such as solar for running clothes dryers would be extravagant, to say the least).

Line-dried clothes are considered fresher and more appealing by outdoor advocates. Considering the clothesline as an appropriate technology is really not so novel for it is a very low-cost device that can result in enormous money savings. However, a major overlooked aspect, besides additional time or inconvenience required to take the clothes outdoors and then gather them in, is the discomfort of exposing certain articles of clothing to public view, either because they are tattered or because they are "unmentionables." Two alternatives are possible for the reluctant outdoor clothes dryer: plant a privacy screen so that the clothes are away from public view; or dry particular items on an indoor rack. Clothes that tend to fade in the sun can be dried inside out or in a hot attic.

CHAPTER 6

Shade Trees and Windbreaks

Trees have been important to us wherever we have lived. We were delighted to get to build our dome home amongst beautiful trees in Uplands Retirement Community in central Tennessee. Maple, oak, walnut, sweet gum, tulip poplar, pine, dogwood, redbud, and others create stunning seasonal pictures from our big windows. Shade trees provide natural air conditioning in summer. Leaves turn exciting colors in the fall, and pines remain on duty to protect us from winter's chilling winds. Fresh greens of spring signal renewed life, bringing new hope and zest to our southern Appalachian lives. Fallen trees even provide some wood for our woodworking shop. Without trees our lives could not be the same!

Rev. Richard and Martha Lammers,
Pleasant Hill, Tennessee

The most natural and aesthetic way to cut cooling and heating costs for Appalachian buildings is to plant and maintain exterior vegetative cover. This involves selection and nurturing of properly placed trees, shrubs, vines, or other dense vegetative growth, as exemplified in the ASPI nature center shown in figure 6.1. Energy-conserving landscaping can play a major role in domestic resource management and is a natural complement to the benefits from overall building insulation and weather stripping around windows and doors. Designing for energy conservation in the landscape is a beneficial extension of our desire to step a little more lightly on the Earth.

Fig. 6.1. ASPI nature center in the shade. The center is located within 500 feet of more than one hundred native woody species.

SITING

The placement of buildings in protected areas away from prevailing winds or with proper solar exposure and accompanying shade trees will add much to domestic comfort levels. This has been known to Appalachians and other eastern Americans from long before the first European settlements. Native Americans were aware that the initial siting is most important; the accompanying trees enhance a site. Appalachians do not like to build homes directly on the top of a mountain. We attempted to convince one group coming to central Appalachia from outside the area not to build on such a scenic location, but our urging was ignored. Unfortunately, two people were killed by lightning on that very site a few years later. Lightning dangers can be extreme on the highest peaks.

Summer temperatures can be as much as 20 degrees cooler in a forest than in nearby open fields. This is due both to natural shading and to evaporation of moisture from the trees, which has its own cooling effect. Heavy-foliage shade trees placed on the southeast, south, and southwest sides of a building to shield it from hot summer sun will have an amazing cooling effect on interior building space and create an interior/exterior atmosphere that is a world of difference away from the rays of a merciless sun. We generally prefer deciduous lawn or border trees, for these trees will shed leaves in autumn and allow the weaker winter sun's rays to still assist in space heating, depending to a large extent on the house's capacity to receive and store such energy.

Bermed soil can have natural cooling or energy-conservation effects when properly placed against the outside walls of the structure. In some respects, a combination of shade trees and berming can work wonders for natural external conservation effects. Wood should not be in contact with the bermed soil.

Windbreaks established on the sides of the structures on which prevailing winter winds strike will have a noticeable effect on conserving the building's heat in winter. The placement of these windbreaks will depend on the topographic features (coves, mountaintops, and other land features) of a particular site, for winds may differ in degree and intensity over only a few hundred feet of elevation on the same mountain. Thus one will have to determine the prevailing winter wind patterns, though in much of middle America normal weather comes from the north, northwest, and west. Since deciduous trees will lose their foliage, the sheltering effect of such vegetation is far less than that of evergreens.

SHADE TREES

Shading is generally more effective for cooling in summer than vegetative barriers are for reducing heating bills in winter. This is in part because some windstorms come from other than the protected direction; breezes quite mischievously change direction and strength. The sun is far more predictable in its path, degree of intensity, and angle of insolation in each season of the year. As is seen in the chapter on passive solar design (chapter 19), the structure itself can be designed and constructed to do much to shield from the fierce summer sun and yet welcome in the winter rays when needed for interior warming. A combination of well-placed deciduous shade trees along with good passive solar design for insulation, ventilation, and protection will equal or surpass the efficacy of artificial air conditioning in many parts of the country.

Placement

Tree planting is key to effective shading and is good appropriate technology design. It is both a utilitarian exercise and a step that makes us more aware of our mortality, for any trees we plant will most likely outlive us by decades. Placement is quite important. Most people are aware that planting trees too close to the building could result in roof and other house damage during windstorms. A certain distance allows for air currents to flow around the house in sultry weather. The same leaves that give us shade in the summer can clog gutters and downspouts without mindful maintenance, but that is not a major determinant in selection and placement. Shade trees placed too close to sidewalks may result in buckling as the trees' root system grows and expands, but that can be partly remedied by changing paths or surface materials. Of course, overly dense trees cannot be allowed to shade solar panels as they will severely impair the solar system's performance. Plant away from gardens, if possible, because tree roots deplete garden nutrients.

One needs to plant trees as one spaces garden vegetables, that is, to visualize what the plant will need in space and size after a certain period of time. Trees grow slowly, but time passes rapidly, so we must anticipate how dense the shading will be and what space each tree will need when more or less mature in one, two, or three decades. Of course, thinning and trimming are possibilities as well.

Selection of Trees

The density of foliage, the height of the tree, and the rate of growth are all factors in selecting trees for shading. Two other factors are use of native trees and edible characteristics of fruit or nuts from the trees. Consider the following:

Black walnuts furnish excellent nuts and have superb-quality wood, but they do not have dense foliage and they disrupt growth cycles for many understory plants.

White walnuts or *butternuts* are also native but are not toxic to understory plants; they give both moderately good foliage and delicious nuts. However, in recent years this species has shown a tendency to succumb to a tree-wasting disease, butternut canker, from which it can be successfully protected.

American chestnuts are making a comeback from a devastating fungus that has been killing off most of the trees in eastern North America since the end of the nineteenth century. New strains are being successfully reintroduced and are furnished by a number of organizations (see chapter 10). This tree is an excellent nut producer, wood furnisher, and shade tree.

Hickories can provide both shade and nuts. Several varieties are native to the eastern United States, where the predominant forests today are oak and hickory. Hickories grow slowly but are quite majestic when given enough room, and their wood is famous for smoking meat.

Oaks come in a variety of species in Appalachia; the pin, red, black, and white oaks are excellent shade trees. The oaks in this part of the country are a principal source of mast for squirrels and other wildlife, and the acorns can be processed into a flour for human consumption.

Elms have been America's favorite shade tree in towns and urban forests. However, many of these majestic trees have succumbed to Dutch elm disease.

Poplars (tulip poplars) are fast growing and are the favorite of many timber people in Appalachia. They grow well in a variety of conditions and elevations, and the flower's nectar is known for the resulting honey. This tree grows into a beautiful shape with dense foliage, and due to its rapid growth it can furnish high-quality shade in a short time.

Ash is another favorite shade variety that has relatively dense foliage and grows tall. It is able to thrive on shallow and rocky soil in Appalachia and nearby regions.

Maple is another popular shade tree that does not seem to be susceptible to Appalachian disease conditions; let's hope this natural resistance will continue. The syrup made from the sap of the sugar maple varieties makes this a favorite for landscaping with edible plants.

Hackberries are well known in the Cumberland Plateau and elsewhere in limestone soils. Though the bark is rough and the wood of little value, the tree has very high wildlife value. The hackberry is a hardy tree that grows fairly rapidly and is a medium shade producer.

Other varieties include beech and birch trees of the Appalachian range, which both furnish good shade and have other benefits as well. On the other end of the size spectrum are native red mulberries, which certainly are not overly tall but do have lower-lying foliage and good fruit for both humans and wildlife. Wild cherries and chokecherries have moderate shade potential and furnish good wood, but they can be deadly when wilted branches are eaten by livestock.

Many people today fill their lawns with decorative trees like the non-fruit-bearing Bartlett pear, which has dense foliage but a slender shape and lovely spring blossoms, as do many ornamental cherries and plums. However, none of these have wide spread and are not substitutes for larger and more sprawling shade trees. Southside decorative evergreens can block valuable winter sun. Some trees that have decorative value, such as the ginkgo (nonnative) and the native black or honey locusts, have a feathery foliage for a more dappled shade.

Those who need quick shade may choose to use succession planting, first planting a variety that grows rapidly but has a relatively short lifetime or breaks more easily in the wind, such as bigtooth aspens. These can fill a space until a slower-growing but sturdier oak begins to yield shade. One should not be tempted to seek quick shade from exotic and invasive species such as the tree of heaven, palownia, Russian olive, bush honeysuckle, or the fast-growing mimosa *(Albizia julibrissin),* but these weed trees may furnish flowers or fruit or give quick vegetation. To fill a niche in the landscape, one may wish to consider planting the backcrossed hybrids of the American chestnut, which are fast growing, hardy, and native.

WIND BARRIERS

Windbreaks have been known for centuries in Europe and Asia and became popular throughout the Midwest and the Great Plains states following the harsh Dust Bowl days of the 1930s. Soil conservation measures were established to retard winds and to conserve topsoil, which could be blown away by prevailing winds. Some of these rows of trees nearer to farm buildings and homesteads are still being used, and they protect the entire farm complex from wind and winter storms. One does not observe many wind barriers in Appalachia. The proper placement of trees will keep soils from eroding and also allow prevailing and chilling winds of winter to pass over or around buildings and reduce overall heat loss.

Trees are planted in clusters or rows perpendicular to the prevailing winds. Such plantings may be arranged as concentric rows of evergreens (such as Canadian hemlock, spruce, or pines) with shorter trees at the outer perimeter and taller trees nearest the buildings. The final barrier should be 40 to 50 feet upwind from a single-story building and include a layer of dense trees with low growth along each edge of the break. Some recommend a central layer of fast-growing trees such as pines. If space is minimal, a single row of spruce trees can make an effective barrier. The evergreen type could vary, and conferring with the local Soil Conservation Service or regional forester is a major help in the final selection.

The exact distance of the barrier from a building depends to some degree on the slope of the hill and the proximity to nearby forests. As with shade trees, too close a planting could allow falling branches to hit the building. Dense windbreaks provide the greatest protection in the area that is two to five times broader than the height of the trees used for the break. The drawback of such a dense break is that negative pressure builds downwind and can draw the air diverted by the barrier back toward the ground. A permeable windbreak allows wind to flow through. Another helpful and energy-saving idea is to place low-growing evergreens close to the foundation of the building for insulation purposes. By preventing wind from whipping around the foundation, evergreen shields can create an air pocket close to the base of the building. In the wintertime, these air pockets can sometimes be as much as 10 degrees warmer than the ambient outdoor temperature.

Evergreens come in many varieties. Appalachia has a good selection of windbreak evergreen candidates, which clothe the hillsides in summer and winter, furnishing beauty and good wildlife cover as well. However, some of these trees suffer from invasive pests and, when stressed by air pollution and other causes, die off and leave a devastated terrain that can be very alarming. In the 1960s the balsam woolly adelgid was imported from Europe and killed many of the Fraser firs in the southern Appalachian mountains at higher elevations. Now a cousin of this pest, the Asian hemlock woolly adelgid, is wreaking havoc with hemlocks. It is coming into Appalachia, threatening to destroy this wonderful lowland/highland evergreen, and particularly moist cove species. The immune systems of Fraser firs and hemlocks have been weakened by air pollution, thus allowing opportunistic insects to cause devastation. Measures may be taken to protect hemlocks, such as the application of insecticidal soaps and the introduction of predator beetles, which have a voracious appetite for adelgids. However, these predator beetles are quite expensive, and the demand for them often outstrips the supply.

In selecting trees for wind barriers, preference should be given to species that have a lot of foliage, grow fast, and are native to the region. One preferable tree is the white pine, which is a true fast-growing variety. Virginia pines also grow dense and well in many parts of Appalachia. Some of the pine varieties are more susceptible to the southern pine beetle, which has so devastated pine forests in the southern Appalachians in recent years. However, it is difficult to select according to possible absence of periodic insect infestation. Some people choose varieties for the proposed arcs of trees and include nonnative species such as the Norwegian spruce, which also grows rapidly and has dense foliage. A possible barrier tree in areas of high limestone soil content is the red cedar, which is widespread in some parts of the region. These do not grow tall for some time, but they have dense foliage when young, are native, and produce a highly rot-resistant wood, which has many good outdoor uses.

LANDSCAPING FOR MULTIPURPOSE USES

The challenge for homesteaders and landscape designers is to select vegetative varieties that can keep Appalachian homes comfortable in both hot summers and cold winters. The region is home to over a hundred native tree species—more than any other temperate forest in the world—and over half of these species would be good for planting near buildings if properly placed and spaced. Many of these are also fruit- and nut-bearing varieties (oaks, hickories, chestnuts) that have a high shade potential. The evergreens can also serve as winter havens for nonmigratory bird species. Choose a healthy mix of hardwood deciduous trees along with evergreen varieties to ward off the winter winds, provide shading, and at the same time serve as wildlife habitat and a source of food for human beings and others. Certain trees are simply better suited for being ornamentals than others. Besides being energy-conservation instruments, the trees can demonstrate the biodiversity and richness of Appalachia, which is part of the beauty of this region.

Shrubbery near a house can protect the building in the winter and also provide food for birds. A bush blueberry can offer a close source of food for the human resident. Vines may serve the same purpose, but people often dislike their tendency to attach to the buildings (causing moisture problems, weakening mortar joints, etc.). However, grapevines may be shaped into a trellis that can serve as a cool resting place from the summer heat.

CHAPTER 7

Food Preservation

As we move toward growing and preserving our own food, we've found micro-climates inside our house to meet a variety of vegetable storage needs. Potatoes, carrots, and cabbages store well in our root cellar; herbs can dry on a suspended window screen in our vented attic; and onions, as we've lately discovered, do best in our unheated back room on an open rack mounted just high enough on the wall to keep from freezing. We learn something every year, and we find ourselves gradually abandoning the ubiquitous supermarket, where so little is asked of us. In preserving our own food, we are leaving a commercial barge for something more like a sailboat that lacks complete predictability, but offers us a more rhythmically complex and delightful life.

Mark Schimmoeller and Jennifer Lindberg,
homesteaders and appropriate technologists,
Franklin County, Kentucky

American gardeners are skilled in growing bountiful crops of nutritious produce, but so often plenty comes all at once in late summer or early autumn. The bounty means giveaways to neighbors, relatives, and the needy—and still there are extra beans, corn, squash, tomatoes, or another crop that was overplanted. Some growers are tempted to follow the example of pragmatic ancestors and squirrel away goodies for winter when fresh produce is scarce. These forebears dried, stored in the ground or root cellars, or preserved foods through smoking, canning, or pickling. In hindsight, we discover that their goals were more than survival; they preserved variety, flavor, and quality for both festivals and ordinary occasions. They developed sophisticated ways to preserve the quality and quantity of their precious produce. We can reacquaint ourselves with their skills and use them together with such modern conveniences as the deep freeze, an appliance they never even dreamed about.

ROOT CELLARS

Root cellars of various designs have been used for centuries and provide convenient, economical storage of unprocessed root crops (potatoes, turnips, carrots), fruit (apples, pears), squash, pumpkins, and leafy vegetables (cabbages, celery). Beyond initial construction, the root cellar is far less resource intensive than canning and freezing methods and does not demand lengthy and often intensive food preparation for storage.

The secret of successful root cellar food preservation is a cool, relatively moist, dark place. The most important factor is climate control. Root cellar temperatures should ideally be between 32 and 40 degrees Fahrenheit, but from 43- to 50-degree short-term storage may be acceptable as well. To insulate the root cellar while providing some ventilation requires an ideal partial submersion using soil as a low-priced insulator. Ideal humidity falls within the range of 80 to 90 percent. Lower humidity causes stored produce to shrivel, and higher humidity results in molding and rotting. Drops in temperature will result in condensation, so insulation (ideally from surrounding earth) is important to maintain constant humidity. Gravel floors are good for maintaining proper humidity. Concrete- and stone-floored cellars may require pans of water to increase humidity, but that is not a major problem in humid Appalachia. Packing root crops in sawdust and sand helps maintain humidity levels as well.

Underground, semiburied, or well-bermed aboveground storage areas can be suitable for food storage. However, the more underground the better. Siting is important for convenience and maximum use of earth; the ideal location is on the north side of a hill. When possible, appropriate technologists use recycled and salvaged materials. They strive to put root cellar footers below the frost line and allow for drainage of excess moisture both from around the walls and from within the floor area to the outside, with exits screened to keep out varmints. They omit cellar windows, for darkness is essential, though the cellar may have an electric light. Some problems can arise with respect to the roof. The roof may be poured concrete reinforced with rebar, or it can be made from sturdy, rot-resistant (black locust or cedar) poles laid close together. Over the roof is placed an 18-inch layer of soil topped with sod, which must be anticipated in the design as a heavy load.

Moving air can be used to adjust temperature and humidity conditions, as well as remove odors and ethylene oxide given off by some ripening produce. Lower-positioned intake and higher-placed exhaust vents are ideal, though they are difficult to build in submerged cellars except at the entrance aperture. Vents are screened to keep out animals and adjustable to regulate airflow. Space is allowed between produce on shelves and all containers to permit maximum circulation of air. Economy-minded builders plan for maximum efficiency and thus design storage areas of the proper size, since it is hard to control unused large rooms.

Some who lack cellars prefer to make storage rooms or compartments in a convenient basement space. These storage areas should be provided with one or more external walls, preferably underground and north-facing, with vents on the external walls. If the walls and ceiling are properly insulated, this interior space could serve as a root cellar. Interior walls need to be rot resistant due to high humidity.

Gardeners learn quite soon not to store onions and garlic in root cellars, for they need dry conditions and can be easily strung in decorative garlands in other rooms in the house. Storing canned vegetables in root cellars is inadvisable because lids may rust. Damaged, bruised, or overripe vegetables are excluded. Excess soil is removed from roots but produce is left unwashed. The stored produce is arranged loosely, with produce needing warmer temperatures toward the top of the room and that requiring lower temperatures at the bottom. People place short-term perishable produce to the front. In very cold weather, operators must occasionally check to keep temperatures above freezing by modifying or closing off ventilation or even adding a heat source.

Root cellars may contain three major classes of produce: quickly perishable produce (green beans, peas, corn, greens); perishables (broccoli, cauliflower, cabbage); and keepers (root crops, pumpkins, squash, celery). Head lettuce, leeks, endive, cauliflower, brussels sprouts, and broccoli should be collected whole and roots inserted in boxes of dirt. Occasional watering should keep them producing for up to four months. Leave keepers in the garden as long as possible. Attempt to segregate fruits from vegetables to avoid odor contamination. If that is not possible, store delicate produce in sand or straw to reduce odor transfer.

CANNING, PICKLING, AND SMOKING

Home-canned fruit, berries, and vegetables generally taste great, but require time and tender loving care by the preserver in the heat of summer. What sacrifices were made for materials and flavor in the past! My mother spent many hot summer days in the kitchen preparing all kinds of things for the winter and was always willing to show off her cellar of canned vegetables, tomato juice, fruit, and berries. It was a beautiful sight to behold and something truly worthy of pride.

Canning is such hot work that many homesteads were fitted with a small, airy outbuilding called "the canning kitchen." A great variety of vegetables were canned, especially the naturally acid ones such as tomatoes, which were put up in a variety of forms (juice, whole tomatoes, ketchup, sauce, and soup mix). Mock mincemeat made from sliced green tomatoes was one of my family's favorite preserved items, for it made great Thanksgiving pies. Other kinds of produce such as corn and beans were mainstays in the canning world, but care had to be taken with nonacidic varieties to prevent botulism through "cold packing" and other canning techniques. In an attempt to conserve food, fuel, and facilities during the Second World War, school cafeterias were opened during summer vacation for growers to can their excess Victory Garden produce.

In Appalachia preserves are foods made with sweeteners, such as jellies, jams, marmalades, fruit butters, and whole-fruit canned goods. At home, my mother made grape, apple, cherry, and blackberry jelly; blackberry jam; plum marmalade (as filling for her famous plum pudding); apple butter; strawberry, tomato, and peach preserves; and canned whole cherries, plums, and berries. Others would add pears to the list. We generally consumed pears raw, for they remained in the root cellar well into the winter. Other folks also made preserves from persimmons, crab apples, elderberries, raspberries, gooseberries, and just about any fruit or berry available.

People skilled in food preservation pickle surplus cucumbers in a variety of ways following home recipes. My mother made sweet, dill, and sour canned pickles from cucumbers as well as brine pickles in 5- and 10-gallon stone jars. She also made pickled watermelon rind. We prepared gallons of sauerkraut in the same large stone jars, and the younger kids would get in after washing their bare feet and trample it until the shredded cabbage was packed tight with its layers of salt. Turnip kraut was my favorite; I haven't even tasted it since those early years at home. The same tasty delight could be made from tender kohlrabi as well as some of the radish crops. We made pickled pigs' feet as well as other pork delicacies such as hogshead cheese and souse made from an assortment of parts (ears, tongue, etc.). For the old-timers in Appalachia, pickling was a principal way to preserve good food for the long winter. But pickling involves plenty of salt, so some people suffered from high blood pressure, a cost of certain food choices.

The smokehouse found on many older Appalachian homesteads was used to great advantage for curing and preserving meat (mainly pork hams, shoulders, bacon, and sausages). The pork pieces were trimmed and dressed to the proper size and shape, salted down, and then hung up to be smoked, usually with hickory wood, for a period of time. The smells in old smokehouses remain pungent decades later and bring back happy memories to old-timers. For Appalachians, smoking was not limited to meat products. Apples were smoked in a ventilated barrel containing smoking wood and a pan of flowers of sulfur, over which the apple slices were hung for a period of time. The stored apples were used for all types of dishes, including "apple stack cake," during the winter months.

SOLAR FOOD DRYING

Solar food drying is one of the oldest methods of preserving foods for long-term storage. Nomadic people have always favored dried food for its light weight (one-quarter of the weight of the fresh produce) and low volume (one-third to one-sixth that of fresh food); ancient Egyptian and Latin American Indian cultures placed dried foods in their tombs. During the world wars dried food became a mainstay of diets in the military. However, in the latter half of the twentieth century dried foods lost popularity to frozen prepackaged and canned foods.

Appalachian people have a long tradition of drying shuck beans, herbal leaves, ginseng root, apples, and other produce with lower moisture content. They often place the gathered, partly dried produce in an old automobile or attic, and in a few days the materials can be stored for winter. In most places, autumn is the ideal time for such drying because the relative humidity is lower, and the produce can reach proper moisture levels for storing before the humidity increases again. Traditional drying spaces have to retain a steady humidity and not fluctuate like outside autumn ambient air during the critical drying period. Drying food in Appalachia is a challenge that requires skill and some care.

Solar food dryers are ideal for use in humid climates where traditional food drying is difficult. These simple nonmechanical devices can be built to require no fossil fuel to operate, and the drying can be done easily and quickly. Sunlight strikes a collector through which a current of air is passing, and the heated air moves over trays of produce in an enclosed screened cabinet (no direct sunlight) and carries off the excess moisture from the produce. The solar dryer can be monitored by shifting the collecting device's direction to follow the sun and by closing the vents at night.

For optimum drying in summer or autumn, the device should be kept out of shaded areas and toward maximum insolation. Some people close vents during autumn nights because of foggy and humid conditions after sunset. Through careful monitoring the foods can be rapidly dried, retaining much of the nutrient content and flavor lost in high-temperature preserving methods. The amount of drying time depends on the moisture content of the produce and on the ambient humidity. Nonelectric solar food dryers can be built by do-it-yourselfers for less than $200 in materials. Figure 7.1 shows how a portion of a solar food dryer can be detached and used as an auxiliary window heater during the cooler portion of the year. ASPI's solar food dryer is shown in figure 7.2.

FREEZING THE SURPLUS

Modern food preservers have access to the deep freeze, which certainly can qualify as an appropriate technology device. The preparation time for a variety of vegetables, fruits, and berries (e.g., asparagus, peas, corn, tomatoes, squash, strawberries, cherries) can be quite short. Produce can be easily blanched,

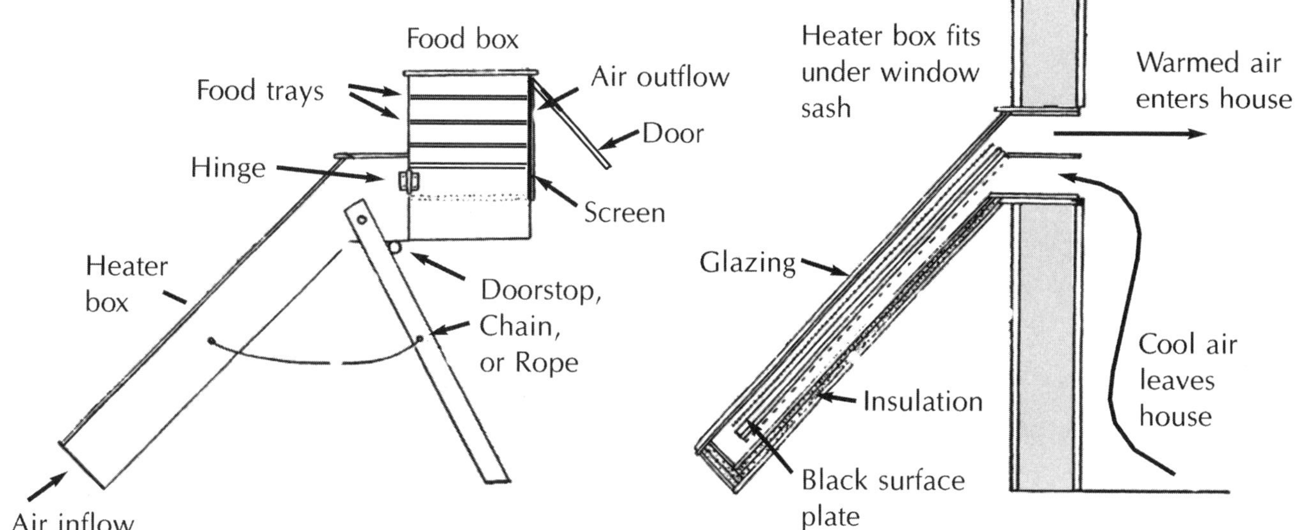

Fig. 7.1. Solar food dryer. The heater box can be detached and used as an auxiliary window heater.

packaged, and stored for winter use. Current utilities, of course, are run largely on nonrenewable fossil fuels, but the use of photovoltaics to run the freezer is a renewable energy option. Several techniques can conserve electricity: buy energy-conserving appliances; open the freezer door as little as possible; keep the freezer full, using ice as filler when food is removed; and clean out the unit in the spring and give it a furlough until the next batch of midsummer produce arrives. This is a good habit because most spring produce (greens, brassicas, and beans) are more difficult to freeze. Deep-freeze space could also be shared with neighbors and relatives.

PROTECTING WITH MULCH AND TEMPORARY COLD FRAMES

The simplest food-preserving technique is to cover the fall crop so that production will continue through autumn and well into

Fig. 7.2. ASPI's solar food dryer

winter or even early spring. Much depends on the type of crop, the microclimate of the garden, the manner in which the crop is preserved, and the weather.

Many root crops can be preserved well into the cold season. In milder Appalachian winters I have preserved carrots, Japanese radishes, turnips, kohlrabi, Jerusalem artichokes, salsify, and even some onions and garlic. The duration depends on the severity of the winter, a good seasonal protective layer, and the appetite of hungry varmints in the vicinity. When under protective cover, hardy greens such as mustard, kale, endive, spinach, and collards will last until late autumn. So will Japanese radishes with their green tops, and even celery, if protected from windburn.

Several methods lend themselves to protecting autumn crops besides taking them indoors to a permanent greenhouse or cold frame (see chapter 11). Fall greens that can withstand some moderate frosts are protected by nightly covering with newspaper or cloths that are removed when the frost melts. A more substantial mulch covering is made from crushed leaves, compost, garden wastes, straw, sawdust, or grass clippings. When lightly covered over to allow leafy vegetables to breathe, the autumn crops will stay fresh and intact for the Christmas holidays and beyond.

Mounds and trenches are other garden-protection techniques for cabbage, celery, and piles of already-dug root crops. Hay bales can be mounded around some vegetables but ventilation is required. Our Appalachian forebears produced "clamps," or earthen storage mounds, located adjacent to a drainage ditch dug to remove excess moisture. A pile of root crops is covered by a 3-foot-tall column of straw; this is sealed with a thin layer of soil, except on the upper end, which is uncovered to help with ventilation. A series of small clamps permits use of only a portion of the crop by uncovering each mound as it is needed.

The temporary cold frame (see chapter 11) is another method of outdoor food preservation. This is a tent, fairly well sealed around the edges, made of plastic or Remay or other fabric. Some cold frames are elongated strips of covering bunched and tied at each end and held down close to the earth by wire pegs. Hoops every 4 feet, made from rebar, wood staves, or curved bamboo poles stuck into the ground, keep the cold frame cover above and not touching the protected produce. This creates a relatively warm and snug microclimate. When possible, the cover is removed to allow the produce to breathe fresh air during the warmer parts of the day.

CHAPTER 8

Edible Landscaping

A hilltop of various wild edible plants is on display and available for taste-testing. . . . Visitors and lodgers are encouraged to roam the countryside. Inspiration Point is one especially interesting place to be on a clear night. A 100-acre forest includes dense woods, a spring with brook and some curious rock outcroppings that people cannot resist climbing.

Narrow Ridge's simple solution to life's mystery is to live in balance and harmony with our environment, to become real inhabitants of the land and to voluntarily choose simplicity whenever possible.

Narrow Ridge Center, Washburn, Tennessee

The lawns around Appalachian homes often look like those in most other parts of America, although they may be smaller, because of limited space. They are decorative but not necessarily beautiful. What is often overlooked is that lawn space can be beautiful when turned into edible landscape. In place of decorative lawns the same land can be used both for recreation and for growing fruit, vegetables, herbs, nuts, and berries. Sitting and conversing is often done on a porch; kids play in the yard.

LANDSCAPE AS DECORATION

Growing a lawn is keeping up with the Joneses, for it is the ultimate sign of conforming within a community. Many affluent people in Appalachia seem to flaunt their wealth in large manicured lawns surrounding their mansions. Aspiring people of modest and lower income then seek to imitate them in decorative landscaping. Those who continue or initiate the practice of edible landscaping become noble resisters of peer pressure. Involved in the current American lawn culture is the colonial mind-set that land is to be set aside for relaxation and enjoyment. The lawn concept stems from Renaissance England's formal gardens and exploded in the twentieth century, through the use of lawn mowers powered by nonrenewable energy and pesticides, to include vast portions of urban and rural domestic yards.

Arguments against such extensive lawns include the time required to mow them; ecological threats from vast quantities of lawn-care chemicals such as pesticides and commercial fertilizers; the need for annual leaf-raking operations; noise and risks associated with lawn cutting, especially on steeper slopes; cost of storing, powering, and maintaining lawn-care equipment; and the air pollutants resulting from using such equipment—a major cause of air pollution in more densely populated areas. Lawns are hard to manage, and thus there is a temptation to use chemical weed killer and to give additional tender loving care to the lawn. In times of drought the lawn made up of nonnative lawn grass is so fragile that it quickly dies, and, to the consternation of the lawn owner, it is generally the first place to be subject to compulsory water restrictions. Lawns are simply too expensive for homeowners, and they can be replaced by productive vegetables. In fact, half of America's fresh produce could come from people's yards in much the manner that vegetables came from Victory Gardens during the Second World War.

EDIBLE LANDSCAPE OPTIONS

Making edible landscapes takes some planning and care. The grower makes a commitment to become ecological, not to be discouraged if neighbors are negative, and to start small and ask those already in the practice for tips and pointers. Often, lawn managers are more confident of lawn mowers and weed killers. To go "edible" is to commit oneself to the Earth while receiving the added bonus of good food as a payment for being ecological. Gardening requires consistent care and nurturing, so convenience is foremost in good planning, especially in choices of perennials such as horseradish or rhubarb. Others turn to low-maintenance fruit and berries because they take less time. Many Appalachians still grow herbs close to the kitchen door as their grandmothers did. The landscaping plans may quickly become involved and include a series of vegetable/flower interplantings or succession plantings producing vegetation all through the growing season and beyond.

Vegetables

A well-managed plot of garden vegetables can be as beautiful as any manicured lawn. When the grower knows what is to be in season, garden greens (spinach, onions, etc.) spring up early, and with forethought the late autumn will be just as green, with kale and mustard and collards; these greens can last well into the Christmas season with the aid of temporary seasonal extenders. Bordering garden plots with perennial plants such as Jerusalem artichokes or purple-flowered salsify plants is one way to extend the garden into the previous lawn gradually, skillfully, successfully. Raised-bed gardening interspersed in lawn areas has proved quite helpful for many who are not totally mobile. Pat Brunner of Berea, Kentucky, is able to maneuver her wheelchair through a flourishing interspersed lawn and flower/vegetable beds. She uses several tools and devices for cultivating and watering without the assistance of others.

Some retirees and others who find bending over burdensome are drawn toward raised-bed gardening both outdoors and in greenhouses. They become experts at intensive gardening, growing a great variety in a small space (see chapter 9). Some prefer to plant vegetables after the dog days of summer and use planters that adorn the external landscape until frost: beets, some beans, cabbage, carrots, celery, Swiss chard, eggplant, endive, garlic, kale, lettuce, mustard, okra, onions, peas, peppers (hot and sweet), radishes, spinach, tomatoes (especially "Tommy toes"), and turnips. Such potted vegetables are brought inside to keep people gardening through the winter.

Fruits

The edible landscape is often way too small to be an orchard, so the cultivator should focus on variety rather than quantity. Growing more varieties allows for several distinctive tastes rather than a cellar full of a few types of produce. Many fruit trees can be selected in dwarf and semidwarf varieties (apple, apricot, peaches, cherries, several plums, and even quinces and pears). Their smaller size means that ladders usually are not needed to harvest the fruit. Selections may be made of heritage types of apples found in Appalachia or of larger standard trees that are planted in corners or borders of property. It is quite possible to add to the landscape other exotic fruits such as kiwis or nectarines and even some semitropical citrus and other fruits growing in pots that can be moved indoors in the colder portions of the year.

Clustering trees together for cross-pollination may be necessary depending on varieties. The self-pollinating trees can be in single stands without requiring others present in the vicinity. North slopes are preferable for fruit trees, especially those most in danger of early blossoming, which could be killed by the late frosts that occur frequently in Appalachia. With care and patience the trees will grow and produce. Nothing is more beautiful than ripening fruit, and birds and certain wildlife (squirrels, possums, and raccoons) generally know it as well. Landscape netting for fruits is now available to keep the birds' share within bounds. Young trees need protection from rabbits and deer.

Grapes

Some of the scenes that I remember best are the clusters of grapes hanging down in the homesteads of relatives. Picking grapes to eat, whether they be wild or tame, brings one closer to the soil. Perhaps it is the power of the vine to pump sap for such long distances that intrigues us so much. Trellises of grapes of a number of varieties (and hardy kiwis) exist in many parts of Appalachia. In some ways, grapevines are the pride of edible landscaping for they grow over trellises or on fencing to form a privacy barrier or to cover patios and backyard sitting areas. The grape canopy overhead furnishes a natural dappled-sunlight effect in the shaded area. Appalachian growers have learned to conquer the mess resulting from falling fruit by stretching under the ripening fruit plastic netting that is slanted away from the gathering or walking areas.

Berries

Appalachians tend to champion a host of varieties and subvarieties of berries, mainly because so many are native to the region and thrive under a variety of soil conditions and at differing elevations. These include blackberries; several varieties of wild and tame raspberries; wild and tame strawberries; high-bush, low-bush, and rabbit-eye blueberries; dewberries; elderberries; and wine berries, gooseberries, and other exotics. Many of these are carefully integrated at Long Branch. Mulberry trees are also strong Appalachian food providers and furnish welcome shade in summer. Cultivated berries are sometimes planted in strips or in beds with corridors of Dutch white clover, which furnishes nitrogen to the berries.

Long Branch uses Blue-XR grow tubes for blueberries, raspberries, and American chestnut seedlings; these tubes provide optimum growth for these and other species. The blue spectrum of light is the most beneficial spectrum for plant growth, and these shelters also amplify blue light, which increases beneficial photosynthetically active radiation. These shelters are also designed to block transmission of a significant amount of harmful ultraviolet light and prevent damage from varmints. An unexpected supplementary benefit is that ladybird beetles or ladybugs (*Coccinellidae* family) breed in the tubes at the base of American chestnut seedlings, so these shelters serve as beneficial insect incubators as well; these beetles are predaceous on aphids, scale insects, mites, and other destructive pests and are used to combat scale in commercial orchards.

Nuts

What is said of fruit can also be said of nuts, though some trees, such as chestnuts and butternuts, have such dense foliage that they become major shade varieties. In Appalachia, people gather a variety of nuts: Hall's hardy almonds; hiccans (a cross between hickory and pecan); butternuts or white walnuts; Persian, English, and black walnuts; hazelnuts; and varieties of hickory nuts. Often these grow in neighboring forests, so that planting the precious little yard space in these varieties is not much favored. One exception is the American chestnut, which almost died out because of a disastrous blight introduced from Asia at the end of the nineteenth century. Today, groups such as Long Branch are trying to reintroduce these species in Appalachia, and every landholding and edible landscape with enough space could benefit from helping to restore this giant to its rightful place in the mountains. (A view of Long Branch's edible landscaping is shown in figure 8.1.)

Flowers

Interspersing flowers or beds of potted plants in lawn areas is common practice. Some people plant petunias, pansies, and impatiens under large trees, thus beautifying ground where little or no grass grows. Large nut trees could be so underplanted, except for walnut trees, which seem to poison the soil beneath the canopy, although certain plants, such as hardy raspberries, are not affected.

Fig. 8.1. Edible landscaping at Long Branch includes raspberries, blueberries, chestnuts, apples, and more.

Flowers serve either to attract friendly insects or to ward off some of the pests. Evening primrose is very high on the pecking order of the Japanese beetle. In fact, it so outdistances other plants in attracting the pests that it can become the collecting point and the insects can then be removed manually without using chemical pesticides. Even without its foliage, the evening primrose goes on to produce highly prized seeds.

Herbs

It is a common practice not only to plant herbs near the kitchen but also to intersperse them in flower beds. Herbs such as parsley and basil provide intense greenery; dill has a beautiful head of seed; and chives have beautiful blooms, which enhance the flower beds' color schemes. The mints add greenery but are so prolific that they demand bordering to control their spread. An added asset is that many herbs serve as spices for Appalachian cooking delights. Herbal tea leaves can replace coffee. Still another advantage is that herbs do not call for the intensive care required for the traditional lawn. Finally, most herbs can be transplanted into pots that can be brought in during the winter season.

Information on site-specific performance, production, insect resistance, drought tolerance, and cold hardiness of fruit trees, berry bushes, vines, and nut trees is available to members of the North American Fruit Explorers (NAFEX) group (www.nafex.org). The group is open to new members.

DOMESTIC WILDSCAPE

Some prefer domestic flowers rather than vegetables or fruits. Wildflowers, preferably native varieties, are highly desired for their aesthetic values. Wildscape is proving to be a living example of low-environmental-impact land use as an alternative to cultivated lawns. In many parts of Appalachia, one sees patches of wildflowers along highway rights-of-way that are cared for by state transportation departments. The wildscape can be an extension of the edible landscape if one takes into account the fact that these wildflowers are food for insects and birds.

There are many reasons for installing a wildscape:

Wildscapes are decorative and yet do not require the chemicals and intensive care of the modern urban lawn.

In contrast to overly managed lawns, the wildscape stands as an inexpensive substitute for commercial seeds, which cost a minimum of 2 cents per square foot. Wildflower seeds are virtually free if gathered by hand in local wildflower patches.

When plants are well chosen, the wildscape can give additional color during all the growing season from April to October.

Growing a wildscape breaks the regimentation of American thought and is a gentle way to convert neighbors to more ecological ways of thinking about the regional landscape and its great floral diversity.

Wildscapes attract beneficial insects and birds. Flowers such as evening primrose, trumpet creeper, butterfly weed, bishop's flower, black-eyed Susan, strawflower, nasturtium, and angelica attract such friends as green lacewings and ladybugs. Some of these flowers attract butterflies as well.

Wildscapes take less water than cultivated landscapes. Though wildscapes are more susceptible to ground fires, dangers are reduced by bordering and walkways.

Wildflowers are our connections with the geologic, genetic, floristic, and cultural ties of our region.

The following suggestions have been gathered through direct experience, by talking with regional wildflower lovers, and from conversations with Gene Wilhelm, author of *Appalachian Highlands: A Field Guide to Ecology,* available through ASPI.

The best defense is an offense. People inform neighbors prior to installing the wildscape so that the neighbors know exactly what is going on. Many potential wildscape growers hesitate because of what the neighbors will think, so bringing them into the planning and selection stage is a way of winning advocates.

Don't let the bureaucrats stop you. Part of changing attitudes is to speak up publicly. One central California wildscape manager was harassed by the fire and police departments. She held her own by refusing to move her wildscape and then gained so many friends that wildscapes began to crop up throughout her block. Even the city departments surrendered.

Border the wildscape. Borders of stone, brick, or even wood enhance the area and give a definitive character to the wildscape. They make a statement that the plot is of value and worth expensive bordering material.

Interplant with trees and berries. Biodiversity adds to the health of the yard, and interplanting with some of the species just mentioned may give an integrating effect to the entire property. Trees should stand at the borders so as not to drain nutrients needed by the wildflowers.

Celebrate the wildflowers in a special way. Some people give a bouquet of wildflowers to a wedding party or for a birthday. Others actually celebrate in the vicinity of the wildscape itself when it is in high bloom.

Promote the landscape. Some have used their best wildflower photos for greeting cards or Christmas cards. Somehow cards mean much more to recipients when the sender grows the wildflowers, takes the photograph, and creates and sends the cards.

BEEKEEPING

The long history of beekeeping, or apiculture, in Appalachia began with the arrival of Europeans. The settlers used hollow logs as bee gums, plugging both ends with removable lids to have access for extracting the honey, and drilling holes for the bees to enter. Smoke was and is used to manage the bee colonies to avoid multiple stings.

The importance of pollinators such as honeybees for vegetables, herbs, and fruits cannot be overstated. At Long Branch as many as thirteen hives have been kept over the years, and copious amounts of honey have been extracted from them. Benefits abound: Buckwheat was planted as a cover crop to enrich organic matter in the soils, and the bees produced generous amounts of buckwheat honey. In the surrounding forest, flowers of the tulip poplar, also known as yellow poplar, produce an almost reddish honey that is quite delectable. Black locusts produce a light-colored and delicious honey. But the most prized of all central and southern Appalachian honeys is the light-colored sourwood honey from the native wild sourwood trees.

There is also a fascinating aspect to beekeeping that still fills the mind with wonder: the dance language of bees, first described by Karl von Frisch. When a female scout bee returns to the hive with pollen and nectar, she performs an elaborate dance for the other forager bees. She first waggles her body from side to side as she moves forward in a straight line, then circles to the right and then to the left, forming a figure eight. This dance clearly communicates to the other forager bees the direction (relative to the sun) and distance from the hive to the nectar source.

Beekeeping is not without its challenges, however. Pests such as Varroa mites, tracheal mites, and wax moths can weaken hives and kill honeybees. Beekeepers must be constantly vigilant to ensure the good health of their hives. But the overall value of pollination and honey production makes beekeeping an extremely important part of successful gardening, edible landscaping, and orcharding.

GREEN LAWN MOWING

We need to be realistic: not all lawn space will be converted to gardens, edible landscape, orchards, or wildscape. For those who prefer a simpler lifestyle approach, one needs to look at lawn mowing practice and methods.

One of the major problems with mowing the lawn in dry weather is that the lawn is cut so short that it will burn out. Lawn services prefer a regular practice (say once every ten days) and do not like to lose business. The tendency is to cut too much and not to tailor mowing to the weather conditions of the growing season. The result is that some weeds can survive better than some of the lawn grass such as Kentucky bluegrass. Overmowing diminishes lawn quality as well as yielding a barren look that lawn owners so dread. They generally compensate by adding precious water (when allowed by local regulations) to keep the lawn green. A higher lawn cut (say 2 inches) will do much more to give a good lawn appearance, for those who value this, than an overly short (1 inch) mowing practice.

If one must have some lawns, we emphasize that steep hillsides or bluffs or overly moist areas are the best candidates for wildscape or ornamental nonmowed ground cover.

Another option is to use a scythe. I used scythes as a youth after my dad instructed me on cutting technique, and I was able to master the rather unique skill. Without some personal instruction most people would be unable to make that tricky device work efficiently. One must hold the scythe parallel to the ground and touch the grass with broad, even strokes. This practice, when applied to grain harvesting or lawn mowing, can be quite efficient and more rapid than one at first surmises. Scythes certainly provide a way to exercise arm and leg muscles and are healthier than riding a power mower; they require no nonrenewable energy, are far less noisy than motorized devices, and can give a wonderful appearance to the lawn when well done.

Reel and human-powered mowers are also available. Some of the newer reel varieties of lawn mowers are low in cost, allow for manual exercise, and yet are far easier to manage than older heavier-duty varieties. (A reel mower includes a set of spiral steel blades that rotate on a horizontal bar set between wheels.) Those older types required muscle, and generally a weaker person found operating them to be quite burdensome. One of the best ways to select a mower is to talk to current owners and to try out their machines on a lawn. If you find a machine that suits your taste, buy one. A reel mower comes at a far lower cost and is perhaps a better investment than a sit-down, powered device—and it's easier on the environment.

CHAPTER 9

Intensive and Organic Gardening and Orcharding

Our mission is to collaboratively create and expand regional community-based and integrated food systems that are locally owned and controlled, environmentally sound, economically viable, and health-promoting.

Our vision is a future food system through the mountains of North Carolina and the Southern Appalachians that provides a safe and nutritious food supply for all segments of society, that is produced, marketed and distributed in a manner that enhances human and environmental health, and that adds economic and social value to rural and urban communities.

Appalachian Sustainable Agriculture Project,
Marshall, North Carolina

Inhabitants of parts of rural Appalachia suffer during winter from lack of fresh, nutritious produce; many urban and rural poor areas also lack sufficient gardening space. In place of healthy food one often finds a high percentage of shelf space in Appalachian food stores devoted to junk food loaded with excess salt, saturated fat, and processed sugar. Obesity, diabetes, and heart disease are epidemic in this and other low-income regions. All too often it is heard that people are "meat and potatoes" folks who never eat salads or fresh produce. State and local governments are beginning to recognize the dangers of junk food, and California has enacted legislation limiting the availability of such food in public schools beginning in 2007. Changing diets is a long educational process and goes counter to the commercial influence of junk and fast food processors and distributors. However, by championing produce from organic gardens and orchards where no commercial pesticides or growth hormones are used and the practice of intensive gardening methods, we can offer the hope of improved nutritional awareness and an expansion of healthy food choices.

INTENSIVE GARDENING AND ORCHARDING

Appalachia is short of large tracts of arable land, and only rarely does one see a mountain plot of land with a larger portion left fallow for the year and a strip plowed for rather extensive cultivation of corn, beans, melons, pumpkins, and squash. Such fields are luxuries seen in some rich river valleys but very rarely on terraced or hillside tracts where gardeners till the same area year after year. While extensive gardening is the rule in the Midwest, where land is more abundant, in Appalachia the key to good gardening is intensive cultivation (see figure 9.1). Steep Appalachian slopes lend themselves to more traditional orchards in various apple-growing districts. Some land-short Appalachians will tend to scatter fruit trees on borders with fencerows, creeks, or roadways to spread the trees over a nonproductive area, rather than taking limited soil for tree crops.

Expert gardeners map out the design for the coming crop year beforehand, but most of us do not like to be constricted by these plans, and we freely change patterns in midyear by removing poor-performing crops or planting more of a certain variety as the spirit moves us. A postmapping is helpful, along with keeping a record of yields of varieties for future reference. Relative position of one vegetable to another is important, as is knowing some types that do not do well in our particular microclimate.

Intensive gardening and new tree plantings take planning, and this becomes a good winter weather exercise for the dedicat-

Fig. 9.1. An intensive garden in Cumberland, Kentucky

ed gardener. What vegetables are preferred, keeping variety, seasonal spread, and spatial needs in mind? Some gardeners prefer beans, corn, tomatoes, and some melons. Other seek about thirty varieties to give diversity. One profitably divides the planned garden into bulk staples that demand the most space (e.g., potatoes, turnips, and tomatoes); produce with medium space requirements based on anticipated need (beans, squash, lettuce, onions, cabbage, cucumbers); and those requiring still less space (herbs for occasional dishes or seasoning). Near-at-hand plots become the herb garden. More distant plots can be ideal places for such perennials as horseradish, rhubarb, strawberries, mint, comfrey, Jerusalem artichokes (on the north side), and cultivated raspberries. Corn is planted only if additional bottomland is available, for it is a heavy user of nutrients, as are sunflowers. Corn can be grown Indian fashion along with hills of beans, which trail up the stalk, with squash for ground cover. Omission of asparagus, which takes rich soil, is suggested where wild poke is harvested; the small poke shoots cook to virtually the same texture and taste as asparagus using identical side ingredients such as garlic. Parsnips take an entire growing season and may be avoided as well.

Even where land is scarce, some gardeners may still prefer to furlough one-seventh of the land (sabbatical) in cover crop and to make sure the late-autumn portion of at least half the land, including tomato areas, is covered with Austrian winter peas, rye, or a noninvasive vetch. The intensive potato and turnip crops can be grown in succession, since potatoes are planted about mid-March and harvested in July and turnips are planted in mid-August and harvested in midautumn. Most tomatoes do not have to be set out until May and so can be interplanted in early cropped areas set out in spinach, radishes, onions, and lettuce. The brassicas (cabbage, broccoli, kale, etc.) should be set out as early as possible and entirely harvested by the start of summer. A hiatus between spring harvest and autumn plantings of brassicas can help discourage insects.

Often greens can be interplanted either before the previous crops are finished or early in the year, and melons and cucumbers can start off within another crop. The early crop can be selectively harvested in concentric circles around the budding late crop as it expands and needs more space as it matures. Likewise, beans can be interplanted with celery, which is a slow grower; the beans will be harvested in midsummer, long before the celery is ready in late summer or autumn. Salsify grows like a grass blade, taking up little space, and can be interplanted with a wide variety of vegetables. Bordering beds with basil and parsley is also a good interplanting technique. Some gardeners prefer not to have beds of one or two vegetables but to have a large number of vegetables interspersed in order to reduce insect damage. Orcharding may be done in much the same manner, with trees interspersed amid cultivated garden plots as well as in a plot set aside only for trees. This works best with dwarf and semidwarf varieties.

Small-scale Appalachian gardeners can grow most if not all their vegetables on intensively cultivated plots or beds of about 4 by 25 foot (100 square feet) minitracts of land. They may even choose their own particular shaped plots. We know gardeners who keep to rural traditions of bringing in a heavy-duty tractor to turn over or "break up" the garden. The clumsy, heavy instrument compacts the plot, and the work is often delayed a month or so due to wet weather conditions. Starting a garden early is a key to success and high yields. Rototillers can be used a little earlier than heavy tractors and cover smaller plots with ease and far less compaction. A third labor-intensive but somewhat appropriate method for early work is not to use power tools at all, only hand tools. The worker stays off the beds to keep the soil loose and does all work from adjacent border paths, typically using spades and digging forks.

Intensive raised beds must be kept loose so that air reaches the roots. Starting to dig a bed is the most difficult task, because double-dug techniques call for spading into the topsoil and turning it over in a furrow. This is followed by digging down a second spade-blade length to the subsoil and loosening it. An easier approach combines raised-bed and double digging:

1. Spade or till soil in a 25 by 4 foot space as deep as possible (preferably 1 foot).

2. Shovel a 1-foot-deep and 2-foot-wide furrow through the length of the tilled bed, piling half the soil in one direction and half in the other.

3. Into the furrow put topsoil from pathways surrounding the bed furrow (about half of a 2-foot-wide path, for the other half of the path topsoil will go to an adjacent bed). The path allows vegetables planted near the edge of the bed to overlap walk space and is not lost space. Now the path floor is about 1 foot lower than the original soil level.

4. Level loose soil evenly over the raised bed. Through the pathway system, one raises the beds of the garden 1 foot

while lowering the surrounding path floor 1 foot. The resulting bed is both raised and double-dug.

5. At the lower end of the raised beds some allowance must be made for natural drainage; add any soil removed in ditching to the raised beds.

6. Add sawdust or cardboard to the paths to raise them to the desired level. A mixture of 4 parts water and 1 part sterile urine can be added to the sawdust to help attain a proper carbon/nitrogen balance. Organic matter placed on paths will turn into compost, which may become soil amendment in a year or so.

An intensive gardener considers both time and space in deciding when and where to plant. Earlier is better, even if late frosts occasionally take their toll. Spreading newspaper or cloth on the young plants on frosty nights allows an early start. My mother always sowed peas in northeast Kentucky before the end of February and had excellent yields. Some of us have a mid-March or first-of-spring launching date for early greens, potatoes, onion bulbs, radishes, and the brassicas. Consider the stair-step approach on a sloped garden with taller plants (okra, staked tomatoes, trellis peas, or pole beans) to the north and smaller plants (squash or spinach) to the south. Many cucumber varieties enjoy being trellised. A beautiful but dangerous plant is the castor bean or mole plant (poisonous to animals and children alike), which is highly effective in keeping burrowing varmints away. Gardeners plant it in parts of the garden away from where children can pick and possibly eat the beans.

ORGANIC PRODUCE

In discussing organic produce we note that the term "organic" as used by chemists refers to most chemical compounds that contain carbon. Originally, organic chemicals were derived or extracted from living substances. Today, most commonly used "organic" chemicals in laboratories and industry are synthetic, including many agrichemicals shunned by organic gardeners. Instead, organic gardeners prefer to emphasize the selection of fertilizing and pest control agents based not on chemical composition as such but on naturally derived sources like compost and manure. They understand that there are immense dangers associated with the use of synthetic pesticides, which are also toxic to humans who may ingest the material during application or as residue on the produce. And simply washing fruits or vegetables with water will not remove oil-soluble chemicals, or all of the water-soluble ones without extra effort.

Scientists have now confirmed that some synthetic agrichemicals in large quantities remain in the soil, resist breaking down to harmless ingredients, and harm beneficial insects as well as other forms of wildlife. Some of these chemicals enter water systems, where they react with chlorinating compounds to produce dangerous derivatives, and are not removed by standard water-treatment processes. In truth, it is best for all concerned to refrain from using such complex synthetic "organic" chemicals unless absolutely necessary. Furthermore, pests become resistant and soils become "addicted" and require increasing doses. The goal is to raise the most nutritious produce in the most environmentally safe way, and chemical pesticides are a dangerous way to go.

There are ways to enhance soils organically. They can be enriched using natural materials such as blood meal, fish byproducts (as used by Native Americans to fertilize corn), seaweed, or other commercially purchased products. However, some of these amendments are expensive. We have tried using diluted urine for nitrogen enrichment along with cover crops of hairy vetch and other nitrogen-fixing legumes. The proper inoculants for maximum nitrogen fixation should be added; these are commercially available, biologically active microorganisms, which are able to increase decomposition of organic materials. We also add domestic compost and decaying leaves to enhance organic matter and spread a thin layer of wood ash (too much would raise the pH to an alkaline level above the generally desired 6.8 to 7.0 range) over the garden for potassium and other minerals found in wood. Others prefer to use green sand as a potassium source. Some additional nutrients and minerals, such as phosphorus (use natural rock phosphates), may have to be added over time if periodic soil tests show the need.

Weed Control

Mulching is valuable in organic gardening. Mulching involves covering surface areas around plants with a thin layer of plastic, wood shavings or chips—not sawdust, for it decomposes too rapidly and takes up needed nitrogen—cardboard, or other inert materials. Leaves can move rapidly to a composted stage suitable for mulch if BioActivator inoculation is added in the autumn (commercial sources are available on the Internet). Hairy vetch is a good vegetative mulch, especially for tomatoes; when vetch foliage is green, the tomato is just beginning to grow; when vetch matures, seeds, and dies back in midsummer, the interplanted tomatoes are nearing their maximum growth. Some prefer to use Dutch white clover *(Trifolium repens)* on paths as a living mulch, but it makes demands for moisture in dry years and can compete with tender garden plants. Mulching with inert materials generally reduces soil temperature, conserves soil looseness and moisture, stimulates earthworm activity, and also retards the growth of weeds. Some find that black plastic raises the temperature of soil too much at certain times and for certain plants, so caution is always advised.

For weed control, chemical herbicides are out of the question in the organic garden. That means more tilling, hoeing, and weeding. A well-worked soil can be looked at on a weekly basis, and even such pests as the invasive European mugwort *(Artemesia vulgaris)* can eventually be controlled. The worst offenders are

Johnson, crab, and wire grasses, which can set in and, if allowed to go uncontrolled, can take over a garden bed and go beyond. Again, attacking the problem early and getting rid of the rooted clumps before the rhizomes spread is absolutely essential. Pigweed, ragweed, and swamp grass can be uprooted early. Raking a bed and letting the sun kill small weeds as they peep through before mulching will diminish their number. Highly nutritious dandelions are tolerated in many Appalachian gardens, for they can furnish a sizable portion of early spring greens—and they require little cultivation!

Pest Control

Chemical pesticides are not used in organic gardening, though it is difficult but not impossible to have commercial-grade, pest-free organic apples. My brother Charlie is approaching a chemical-free orchard by using selected disease-free varieties and certain natural pest-control agents. No pesticides are used at the orchards of Long Branch, but efforts are directed to pressing fresh cider, which can be made from cosmetically imperfect fruit—products uninformed consumers might reject in favor of fruit with perfect peels covered with invisible toxic chemicals.

Animals can be allies of organic gardeners. Early homesteaders allowed the geese to wander in certain areas of the garden because they would pick off the insects. Many birds, such as robins, are immediately attracted to the organic garden and will assist in pest control. Bringing in beneficial insects (praying mantises and the ever-more-numerous ladybugs) and parasitoids (parasites that kill their host) such as wasps can also be effective. Other biologicals include bacteria for caterpillars *(Bacillus thuringiensis)* and the milky spore for Japanese beetle grubs.

Certain plants are also useful in the organic garden. We mentioned castor or mole beans. Certain flowers, such as marigolds and others suggested in the sizable organic literature, may be interplanted in the garden to ward off various pests. Many natural pesticides are made from plants or are botanical controls that break down quickly in the environment through natural biological activity. These include pyrethrin, rotenone, ryania, and sabadilla. The last is a natural broad-spectrum insect killer made from the crushed seeds of a lilylike Caribbean plant and is used to kill everything from cabbage worms and cucumber beetles to citrus thrips and leafhoppers.

Crop rotation can also help in pest control. A general principle is to rotate cropping and not raise the same susceptible vegetables two years in a row in the same bed, or to refrain from raising them at all for a year to retard the pests.

Some large insects like tomato and tobacco worms can be removed and killed manually as we did in tobacco fields when I was a kid. Another that can be removed manually, even when caring for a hundred tomato plants, is the Colorado potato bug. Smashing clusters of early eggs on squash leaves can reduce the number of squash bugs immensely. As mentioned elsewhere, we find that evening primrose is a favorite host plant for the Japanese beetle; these bugs cluster on the blooms and can easily be shaken off into a container of soapy water.

A number of lures and traps are available that are specific for certain pests and are biodegradable. However, effective lures can attract all the pests from miles around, which is not necessarily the result an individual gardener wants. Safer's soaps, which will kill certain insects as well as powdery mildew, moss, and algae, are selective, safe, organic, and biodegradable and can be obtained locally or through catalogs.

Intensive gardening is not new to Appalachia; nor is organic gardening, for it was the only way until the middle of the twentieth century. These methods have generally gone out of popularity with the abandonment of gardening and the turning of literally billions of food dollars (which could be saved through gardening) over to the fast food giants. It is time to regain control of our food supply. Returning to safe, available, low-priced, fresh, organic food from garden and orchard is a goal that may be a long way off, but the rising popularity of organic produce is bringing back good nutrition to poorer regions as well as affluent ones. For the health of our people it cannot come too soon.

CHAPTER 10

Regional Heritage Plants

> *In the United States the name of the cooperative effort (to preserve species) between the Bureau of Land Management (BLM) and Kew Royal Botanical Gardens is called "Seeds of Success." To collect the wild plant seed the BLM in many cases has contracted the work of young botanists through the Student Conservation Association. I was one of those young botanists. In the summer of 2005, we roamed the mountains and valleys of southeastern Idaho in a grand search of plant species. Many times we found a promising flowering population of a plant that had not yet been collected. Then several weeks later when the flower had matured into seed we returned to collect the seed. In this manner our small team collected seeds from about 70 plant species.*
>
> John Thomas, biology student,
> Goshen College, Goshen, Indiana

It is no accident that the Cherokee refer to the southern Appalachian mountains as "the birthplace of all the plant people." The genetic material passed on in plants and animals through generations is an evolutionary heritage, a natural equivalent to sacred songs, ceremonial dances, and the solemn passing down of cultural traditions on the human level. Nature in an unheralded manner passes on its tradition through a tortuous process of sorting out living forms that could survive under specific environmental conditions of soil, climate, and unusual weather conditions. The wild animals (see chapter 12) and plants have survived and thrived provided there was no major human or other interference. The land sections of this book include information on restoring damaged land using healthy plant species. Here, our attention is given to plants that are also part of our food chain (nuts, fruits, herbs, and vegetables); when that chain is weakened, our very survival is at stake.

Many food plants have been threatened or endangered by monocultural practices in agriculture. It is imperative to question short-term practices that advocate growing certain plants from seeds that are mass produced, using genetically altered plants that can withstand certain pest assaults or weather conditions—and that prove more profitable to corporate interests. We soon forget that these cultivated popular species may have hidden weaknesses that will only come to light when it is too late. Furthermore, these species may not withstand unexpected threats and may be wiped out, as Irish potatoes were in 1848.

Biodiversity protection and continuity are crucial for the Earth's health and our human survival. Reducing that diversity, as is being done by modern seed-propagation techniques and business practices, may ultimately be harmful to the entire biosphere. Good Earth insurance includes plant protection directed at enhancement of our global genetic wealth.

Part of healing the Earth is restoring this threatened heritage of native species. We could talk about the answers actual and possible to a host of threats, from rain forests being cut to coral reefs being damaged, from prairie grasses being plowed or trampled to desert flowers being run down by dune buggies. But it is best to restore some food plants well with the strong resolve that the movement will grow to include all plants whether presently useful or not, for biodiversity covers an immense range. We include here four special topics associated with regional heritage concerns: American chestnuts; apples, the grand temperate fruit, including crab apples and exotic varieties; heritage herbs; and seed-saving techniques.

AMERICAN CHESTNUTS

> *They also grew behind my house, and one large tree, which almost overshadowed it, was, when in flower, a bouquet which scented the whole neighborhood, but the squirrels and the jays got the most of its fruit; the last coming in flocks early in the morning and picking the nuts out of the burs before they fell.*
>
> Henry David Thoreau, *Walden* (1854)

I am too young to remember the blight that killed our chestnuts *(Castanea americanus),* but I recall a massive 3-foot-diameter ghost trunk on our farm's border. We kids brought a chunk into the shop where my dad was working; he told us to take it back, as though we had disturbed a grave. My dad kept the farm quite tidy, but he left the chestnut's grayish white remains for years just where they fell—and we never forgot his hurt.

In 1904, a fungus disease, *Cryphonectria* (formerly *Endothia*) *parasitica,* believed to have been imported on Asian chestnuts in the 1880s, reached the New York Botanical Gardens. For the next half century it continued a slow progress of destruction that killed millions of exemplars of the grandest tree in the eastern United States. Few remain today; although the roots were not affected and continued to produce sprouts for years, many if not most of the young trees became infected before reaching maturity.

Early efforts to develop a blight-resistant strain of pure American chestnut failed and were all but abandoned by the mid-twentieth century. However, researchers found a weaker form of chestnut blight fungus that consists of a virus that is transmitted to virulent strains of the disease under certain conditions and that allows the trees attacked by these strains to survive (hypovirulence). A second method is to backcross American and Chinese chestnut hybrids to breed for resistant chestnuts and by this process to develop a forest from trees with blight resistance. A third method is to obtain resistant chestnuts by transferring to them resistant genes.

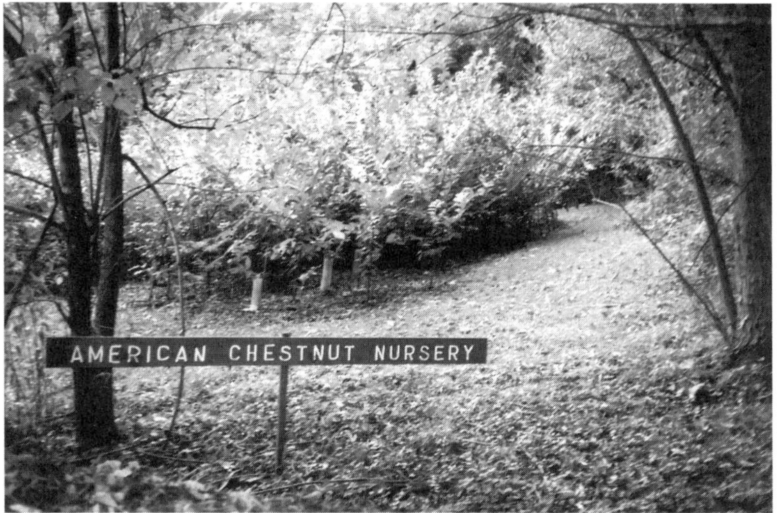

Fig. 10.1. This nursery is part of the American Chestnut Restoration project at Long Branch.

Several groups work on this (the American Chestnut Foundation, the American Chestnut Cooperators' Foundation, and the American Chestnut Regeneration Effort). The American Chestnut Restoration project at Long Branch has as its mission to return the American chestnut to its niche as a provider of food and habitat for wildlife in the southern Appalachians (see figure 10.1). The Chestnut Restoration project grows and distributes seedlings to interested growers for forest restoration and edible landscaping and regards each act of planting trees as an act of hope for the future. The project uses a backcross of the American chestnut with the Chinese chestnut *(Castanea mollissima).* One of the center's buildings is an early twentieth-century traditional mountain farmhouse made and furnished with local, wormy American chestnut wood.

People are seeking out and discovering American chestnut survivors; they are planting new stands of chestnuts in pairs; they are encouraging local conservation groups and national and state parks to embark on major chestnut-restoration efforts. Truly, this is a task worthy of American Earth healers.

HEIRLOOM APPLES

Long before Johnny Appleseed (John Chapman) sowed apple seeds in the Midwest and planted frontier nurseries, apples *(Pyrus malus)* were important in Eurasia and (as crab apples) in North America. They were domesticated about 6500 B.C. and have been the grand Northern Hemisphere temperate fruit ever since. A member of the rose family (Rosaceae), the gently fragrant and delicate apple blossoms are harbingers of spring. What we do not often realize is that many varieties have escaped public notice as most commercial apples are of only about a dozen varieties, which are in perfect form but are heavily indebted to chemical pesticides to make them cosmetically marketable. And washing with a little water does not remove all pesticides.

Greeks, Etruscans, and Romans all practiced grafting favorite apples onto other apple and crab apple rootstock. Since apples are heterozygous and thus do not reproduce true to type, saving and planting seeds from a favorite apple tree will not always yield trees of that variety. Apple trees are reproduced by grafting or budding. By these methods, a scion, or bud of the desired variety, is inserted into the base of the stem or trunk of a seedling tree known as the stock, and sometimes the stock itself is a vegetatively propagated tree. Today, improved varieties include trees of uniform moderate heights that minimize the need for ladders during harvest. Extremely dwarfed rootstocks are popular where space is limited, and espaliers can make picking very convenient as well as create beauty in an edible landscape.

Heirlooms have their own particular flavors and growing habits. They often have unusually long or late seasons that allow them to elude late spring frosts and freezes. They have a natural resistance to pests and diseases, and thus growing them can minimize the use of harmful agrichemicals against the plagues of the apple world, namely, apple scabs, mildew and fire blight, coddling moths, apple maggots, red-banded leaf rollers, aphids, leafhoppers, mites, and the like. Heirlooms are often but not always better candidates for the organic apple, something still quite difficult to achieve.

The region of Appalachia alone once contained 1,300 known varieties of apples, of which only a few have been preserved. Among them some of the better known are the York Imperial, Delicious, and Stayman Winesap. In the North Carolina mountains 90 percent of commercial production consists of Red Delicious, Golden Delicious, and Rome varieties. People have forgotten the many varieties of heirlooms that made the best apple pie, cider, applejack, vinegar, apple stack cake, and apple butter. Just listing these heirloom varieties would fill this book: Arkansas Black, Bushy Mountain Limbertwig, Crow's Egg, Grimes Golden, Hoover, Kinnaird's Choice, Magnum Bonum, Newtown Pippin, Russet, and on and on. New varieties, which are disease resistant, are also emerging, including my Newark, Ohio, orcharding brother's apple varieties. These add to the apple heritage list. While all apple lovers welcome new additions with unique flavors and cooking or preserving qualities, apple-variety conservationists are constantly on the lookout for old heirloom apples from which cuttings can be taken to continue heritage apples here in the mountains.

HERITAGE HERBS

Growing herbs has been a tradition for centuries among those seeking good flavors, well-seasoned cooking, and pleasant-tasting teas. Herb growers learn their skills from parents, other relatives, and friends. In rural America most homesteaders grew perennial herbs such as sage *(Salvia officinalis)* and chives near the kitchen door; in plots farther away they grew annuals like parsley *(Petroselinum crispum)*, basil *(Ocimum basilicum)*, and dill *(Anethum graveolens)*. While commercial dried spices and seasonings were purchased, local fresh herbs were even more appreciated. More-accomplished herb growers would cultivate native varieties such as Appalachian perennials wild bergamot *(Monarda fistulosa)* or white bee balm *(M. clinopodia)*. Controlled exotics were either of European origin—oregano (*Origanum* spp.) and rosemary *(Rosmarinus officinalis)*—or natives like mountain mint *(Pycnanthemum incanum)*. Uncontrolled exotics, such as mints (*Mentha* spp.), needed borders to prevent their spread.

Today's urban dwellers have more restricted time and space for growing herbs. Meals are often more rapidly prepared, and popular seasonings (chili, mustard, and ketchup) predominate. Fresh herbs lose their appeal, and some varieties, like fenugreek *(Trigonella foenum-graecum)* and dittany *(Cunila origanoides)*, fall into disuse. If these herbal names seem foreign, they should not, for both are among the oldest and most widely used of native herbs. Many plants that were well known in the eighteenth or nineteenth century are little known today except in herbalist circles, and they are rapidly moving from common to heritage varieties. Their restricted use makes them candidates for the museum shelf of botanicals. But with renewed interest on the part of heritage-herb growers they can be restored to their rightful place of honor and respect.

Herbs offer a variety of flavors and smells, making those who grow them more respectful of the richness of nature. Growing herbs offers several advantages: They are easy to cultivate, because they are hardy and do not require soil tilling or soil amendments; they are healthful and nontoxic (although in rare cases, such as comfrey, they could be overused); they are easy to store and dry; and they can form some of the most beautiful aspects of an edible landscape. As people come to appreciate them, the market for them is growing. In addition, some herbs can serve as substitutes for coffee and other caffeinated drinks; being home-grown, they reduce food bills; and they are part of the plant heritage.

Perennial herbs do not require plowing and replanting each year, nor do they need automated horticultural care. Being closer genetically to wild plants, herbs are more resistant to insects and diseases; some even repel or discourage insect pests and serve as their own weed control. Perennials require less water and fewer nutrients than conventional crops. Processing herbs necessitates hand labor and thus provides a basis for local cottage industries.

SEED-SAVING TECHNIQUES

Open-pollinated seed varieties are becoming scarce, so that we are now in danger of losing a priceless heritage. Organic growers have a mission to do more than preserve varieties: seed-saving saves money, allows for broader choice, and perpetuates plants that thrive in particular conditions. In my youth, we saved many vegetable and tobacco seeds. During the mid-twentieth century we saved less and less as hybrids took over, seed prices were reasonable, and saving seeds seemed to be a greater bother. The seed catalog replaced our seed-saving habits, but we were losing something precious. We had for years saved seed from a juicy yellow-to-pink tomato that withstood drought. We lost the seed over time and forgot to mourn. However, we are now beginning to recognize the critical importance of seed saving, and many regional and watershed seed-saving groups are springing up.

All Earth-healing processes take certain skills, and growing seeds is no exception. Plant reproduction requires knowledge of plant parts. Specific seed-saving techniques apply to each cultivar. Is the plant a hybrid? An annual, perennial (like rhubarb or asparagus), or biennial (which flowers and produces seeds the second year)? Is seed self-pollinated, or are there male and female plants? How susceptible are plants to cross-pollination, or how far apart must the cultivar of the same species or genus be placed to minimize cross-pollination (for vegetables, the distance can vary from 100 feet to a mile for pure seed)? Are plants isolated by time (planting at different intervals) or by mechanical means (cages or floating row covers that require hand pollination)?

Seed may be saved for personal use, to exchange with others, or to start a business to breed plants for specific characteristics: earliness, vigor, color, or resistance to disease, insects, or drought. Seeds are saved from the healthiest plants that exhibit the desired characteristic—thus the seed collector becomes involved in the evolutionary process. Harvesting time must be after maturing and just when the first seed begins to drop. Some seeds ripen over time and are saved in plastic covers. Clustering seeds (e.g., salsify) can be easily collected. Some (such as legumes) are picked from dry vines; some grains are threshed; and some (e.g., corn) are shelled. Tomato seeds can be collected by squeezing tomatoes through a strainer or collecting the heavier seeds that settle to the bottom in tomato slurry fermentation; these seeds are air-dried on newspaper and then put in airtight storage containers.

Efforts are being made by public and private groups and individual volunteers to bank saved seeds. The Millennium Seed Bank started by the United Kingdom's Royal Botanic Gardens, Kew, intends to bank 10 percent of the world's flora seeds, or 24,000 species, by 2010. Currently the seed bank is working with the Chicago Botanic Garden to save 1,500 tall-grass prairie species in the Midwest, where 9 species are already endangered and the native prairie is shrinking to small patches. One danger is that if the seed is in a bank, people may have a sense of false security that it is protected—but seeds are more valuable than gold bullion, and the life force in the seeds can ultimately only be realized by allowing them to grow and reproduce. Seed banks are critically important tools for protecting a precious legacy.

Perhaps none of the methods discussed in this book involves working more closely with nature than saving heritage varieties. The techniques mentioned here are basic and only scratch the surface; methods quickly get more sophisticated. But the key is to realize the importance of saving heritage species and to encourage those with botanical curiosity and a desire to play an important role in plant evolution. Prior skills help, but natural interest and curiosity can stimulate an investigation into a host of helpful sources: county agents, extension services, professional growers, Internet sites, libraries, greenhouses, and other places.

CHAPTER 11

Solar Greenhouses and Season Extenders

From both a mythological and technological point of view, the use of glass in solar greenhouses and sunspace building design represents a significant convergence of both overarching aspiration to reaffirm a connection with the infinite, and on-the-ground practice of creating sustainable relationships. It marks a major turning point in redefining the role of human shelter design and the Earth.

Of course, the ancient ones, the Ancestral Puebloans (Anasazi), oriented their cliff dwellings and their ceremonial centers such as Chaco Canyon to the movement of the sun in winter. But now glass not only gives us protection from the wind, rain, snow, and subfreezing temperatures, it also allows us to become a sentient part of the outdoor landscape, with wildlife wandering by, falling leaves, passing clouds, the flight of birds, rainbows, sundogs, thunder, lightning, and even the jeweled net of the sky dome and Grandmother Moon!

The inner world of our built environment is now inextricably linked with the vast beauty and harmony of nature! And so we're mightily inspired to build and grow with the sun and the green.

Paul Gallimore, Long Branch
Environmental Education Center,

Solar greenhouses are fascinating models of energy and food self-reliance in different parts of Appalachia, merging food production and heat energy production into aesthetically pleasing architectural creations. When attached to existing buildings, these greenhouses are able to share their surplus heat during winter months with the adjacent space. Thus they have the potential to convert these combined structures into functional bioshelters that share warmth, oxygen, and food.

Appalachia has a growing number of greenhouses and other types of seasonal extenders (created environments for protecting plants at each end of the growing season). These are used mainly for growing produce and flowers but have many other benefits as well. However, some traditional greenhouses are poorly insulated, use expensive and sometimes scarce propane fuel, and require full productive output of the commercial crop just to pay for themselves. Recently, one greenhouse owner tried to sell his supposedly thriving business to one of us, but when pressed, he reluctantly confessed that the propane prices were running him out of business. Solar adaptations were suggested, but the business owner said he was unfamiliar with solar applications. Many people in the region still lack an understanding of the basic principles of efficient solar design and how these techniques can be used in greenhouse design and construction.

Solar greenhouses and other permanent seasonal extenders are quite simple, utilizing the basic principles of collecting, storing, and distributing the sun's energy (see figure 11.1). Shortwave solar radiation penetrates the greenhouse's transparent glazing and, in part, is converted to longer-wave heat energy within the enclosure (as in cars parked in sunlight in winter). The longer-wave energy finds it harder to escape, thus raising interior temperatures. Dark-colored surfaces absorb this heat and allow more of it to be retained by certain types of dense materials (stone, masonry walls, and especially water stored in dark containers). Plants benefit from this warmth during cooler nights as the heat-storing materials gradually emit their warmth into the cooler greenhouse environment.

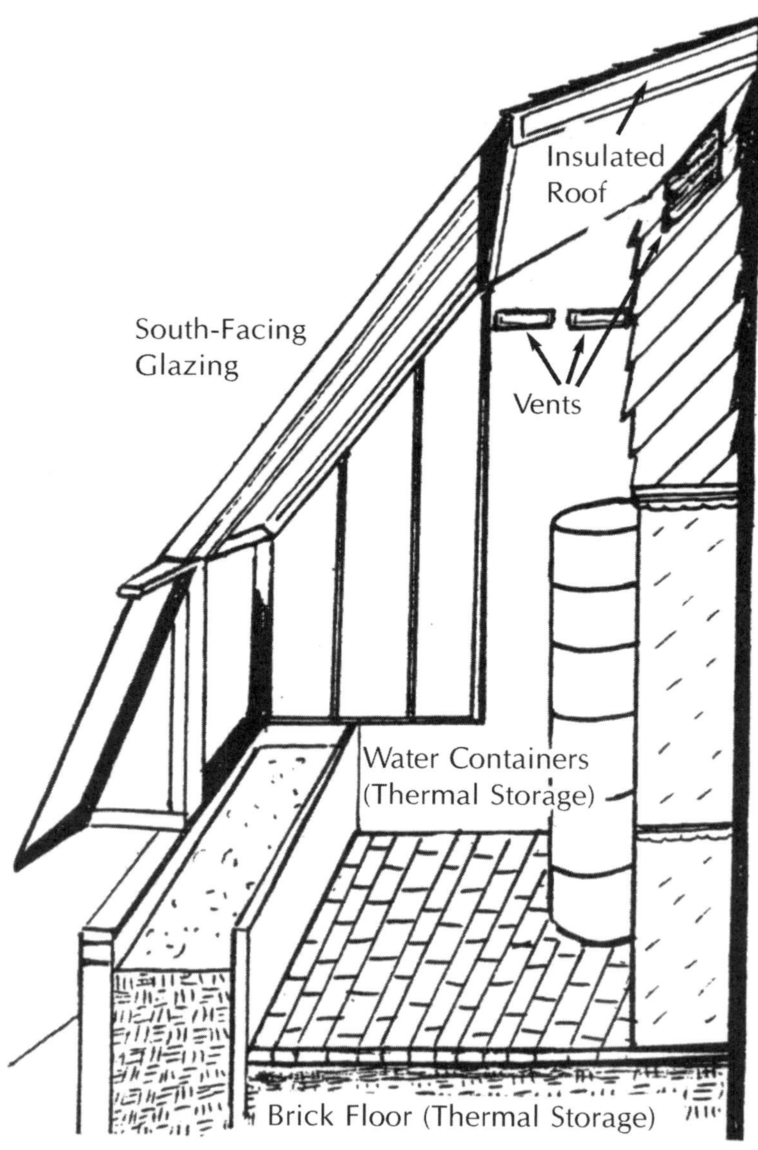

Fig. 11.1. Attached solar greenhouse

A properly sited, constructed, maintained, and utilized solar greenhouse or seasonal extender is perhaps the most versatile form of appropriate technology and provides the following:

a ready source of nutritious food, especially greens and brassicas, during the winter months when fresh produce must be shipped in from great distances and is quite expensive

an excellent opportunity for getting moderate exercise through horticultural activities in winter, even for the disabled. It allows people to keep in touch with the soil—a psychologically healthy enterprise at any time of year, but especially so during the cold winter months.

when large enough, an aesthetically pleasing living space for meditation and rest. The Long Branch Center's largest solar greenhouse includes a solar composting toilet and flourishes with an abundance of fragrant flowers, vegetables, and vines (see figure 11.2).

Fig. 11.2. Solar greenhouse at Long Branch

When attached to another building, a solar greenhouse used as an air lock can reduce cold-air exchange with the wintry outdoors. Adequately sized greenhouses can furnish a portion of the attached building's space heating. The 200-square-foot attached solar greenhouse at ASPI's office in Mount Vernon, Kentucky, supplies 40 percent of the space heating needed during the winter months by a 2,000-square-foot office space.

Solar greenhouses and other seasonal extenders, as with any appropriate technology, require careful attention to basic design and operation. They should be sited with a good south-facing aspect; be tightly constructed with superior insulation; and be well maintained. And, of course, it is important to choose the right vegetables, containerized plants, and herbs to grow.

SITING AND PLACEMENT

Orienting the south face of the greenhouse 30 degrees east or west of due south is good placement, sacrificing only 8 percent of efficiency. Although it is somewhat easier to incorporate an attached greenhouse into the design of a totally new construction project, with their great versatility greenhouses can be retrofitted onto existing buildings.

In the winter, the greenhouse may be shaded because of topography, other buildings, or evergreen trees. It is helpful to obtain a Solar Pathfinder, which can indicate how much the sun will strike a given site throughout the year (www.solarpathfinder.com).

Inserting a seasonal extender into the side of a south-facing slope furnishes good low-cost insulation provided the glazing faces south or southwest. Sometimes the ideal south-facing direction may have to be sacrificed due to limited possible locations. Some greenhouse plants grow best with morning (southeast) or evening (southwest) sun, though more eastward and westward locations will require additional heat storage, shading, and ventilation. Sometimes the best site is on the roof or attached to the side of a building with some adjustment for easier access. In some cases an evergreen tree may have to be trimmed or removed, but deciduous trees are desirable summer shade assets.

It is hard to imagine an attached solar greenhouse that is too large. The interior space can yield an abundance of fresh winter vegetables, a number of seedling flats ready for transplanting into outdoor gardens, and ornamental flowers, shrubs, vines, and even small trees, which can lend the air of a conservatory or botanical garden to the space. With enough room, solar composting toilets, aquaculture tanks, and hot tubs can provide for multifunctional use of the space. For plant lovers and solar aficionados, no solar greenhouse space will go underutilized or unappreciated.

If the surrounding landscape permits, the prevailing winter winds can be diverted from directly striking the solar greenhouse by evergreen windbreaks (see chapter 6). During the summer months when greenhouses without sufficient thermal mass are prone to overheating, a useful annual native plant, the North American Jerusalem artichoke *(Helianthus tuberosus)*, may be grown in the soil or in a series of containers in front of the greenhouse glazing for shading. By midsummer, the plants can grow as tall as 12 feet and provide abundant vegetative growth to shade out the fierce summer sun. The foliage thins with the coming of the flowers by early autumn, when the sun is less intense. Of course, the added benefit is that Jerusalem artichoke tubers furnish an excellent cooked vegetable or delicious garnish for salads all through the winter. For greenhouses that are two or more stories tall, deciduous fruit or nut trees provide exquisite summer shade and food for wildlife and allow the winter sun to heat the greenhouse.

Architectural beauty aside, the design of the seasonal extender should fit one's needs, building skills, and financial resources. Needs are relative, and the perfect size may often be smaller than originally thought, because larger greenhouses take more skill to build; building and gardening skills are generally acquired over time and through neighbors with expertise; the financial investment is generally small and the payback is quite short, usually only a few years. A final design note: an attached greenhouse will cost less than a freestanding structure and will provide heat for the attached space.

CONSTRUCTION

Appalachia abounds with native stone, sand, gravel for paths and wall construction, wood for framing and siding, and native river cane for trellises. Plentiful native black locust *(Robinia pseudoacacia)* and eastern red cedar *(Juniperus virginiana)* wood can be used in the greenhouse because of their natural rot and decay resistance in moist conditions. By using these woods, the toxicity of chromated copper arsenate (CCA) pressure-treated wood can be avoided. Flower and produce containers of clay or plastic resist rot; growing beds bordered with rock will never decompose; and shelving can be set on top of water tanks or barrels.

The common building material, fiber-reinforced cement board, sold under the trade name of Durock Brand Cement Board, will not swell, soften, delaminate or disintegrate; it is also noncombustible and carries a thirty-year limited warranty. Although it is typically used as a ceramic tile backer board to provide a water-durable base for walls, ceilings, and countertops—anywhere tile is used—this material has proven highly successful in one of Long Branch's solar greenhouses as a barrier to hold soil in the planting beds and also as shelving material for supporting potted plants and seedling flats. Its nondestructible characteristics make it ideal in the high temperatures and high humidity of greenhouses, and its density helps to provide some thermal storage for the building.

Finding proper glazing materials in parts of Appalachia can sometimes be challenging. The glazing material used in the Long Branch greenhouses is salvaged tempered glass combined with readily available insulated sliding glass door units. Salvaged double-hung windows are used in the east and west walls to allow cross-flow ventilation. The advantages of tempered glass over plastic and various composite materials are cost savings and durability. Plastics will degrade over time in ultraviolet light and must be periodically replaced. Although the initial costs of plastic are less, the labor and glazing replacement costs will be much higher than with glass. The other great advantage of tempered glass is that it does not obscure extraordinary mountain views, and it allows one to experience the great outdoors, the seasons, clouds, weather, passing wildlife, and birds while still being in a subtropical, protected space.

Having noted the advantages of glass, something can be said for certain plastic glazings. The ASPI office greenhouse is covered with Lexan, a product manufactured less than 200 miles away by the Pierce-Ohio Companies. While the ⅜-inch, two-wall, corrugated polycarbonate Lexan sheets are translucent, the material is strong and virtually unbreakable, and its light weight makes it easy to ship, cut to shape, and install. While lacking the clarity of glass, the polycarbonate diffuses the sun's rays, which is excellent for plant growth. Some may argue that nonrenewable resources were used in the manufacture of Lexan, but glass, too, has considerable "embodied" or process energy content. After a decade there is no sign of deterioration from UV exposure, and the Lexan has withstood a hailstorm with no trace of damage.

Efficient builders often realize substantial cost savings by designing their greenhouses around windows or glazing materials already on hand, rather than purchasing new materials to fit a static design. Framing made from native black locust and eastern red cedar wood is the most durable. When using salvaged glass to construct thermal pane units on site, the use of weep holes helps to keep the glazing units from fogging up due to condensation. Single-pane glazing will allow too much heat to escape and is not recommended. Interior wood surfaces can be treated with half-and-half linseed oil and turpentine with a bit of paraffin added.

A light-colored roof will help to keep the building a little cooler in summer. Both roof and glazing slopes have proven important for the efficiency of the collecting structure, and this is especially true in steep terrain. Vertical glazing for greenhouses is practically leakproof and allows for minimal reflection of winter sunlight. Angles of less than 58 degrees from horizonal will cause excessive reflection of the winter sun.

For maximum efficiency and heat retention, the foundation and outer walls of the greenhouse must be properly insulated. Stone, brick, block, or other insulated masonry walls along with water (stored in dark blue containers—barrels, drums, or even plastic milk jugs) are good and readily available heat-retaining materials. (See chapter 8 on blue containers.)

In larger greenhouses, fish tanks can be installed that serve a dual purpose: they can both retain heat and grow fish. Careful attention to caulking and insulating materials is essential, for the less outside air infiltrates and the better the insulation, the better the greenhouse performance. Season extenders can be dug into a south-facing slope using the earth itself as an insulator; these are akin to "pit-dug" seasonal extenders, simple trenches covered with glazing, found in parts of the South.

For plant watering, rainwater collected from the greenhouse or attached building roof is far superior to chlorinated municipal water. A solar photovoltaic array can provide light for the greenhouse and for warming seed mats in late winter. Heating of soil sufficient for germinating spring plants is not difficult in a solar greenhouse. Some brassicas and even celery can be germinated in soil beds rather than in individual trays.

MAINTENANCE

Greenhouse operators know that ventilation is a key to good greenhouse functioning in order to discourage some of the damping-off funguses. Figure 11.2 shows vent placement to maximize fresh-air intake and hot-air emission. Such a vent should be movable. A thermostatically controlled small exhaust fan and a vent placed on opposite walls will allow air to flow

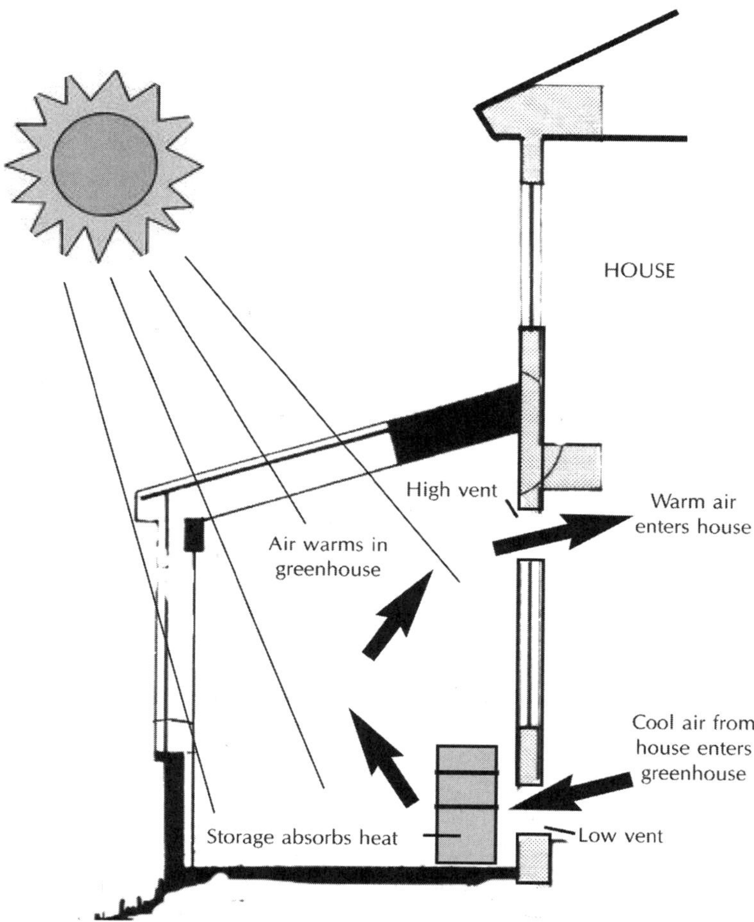

Fig. 11.3. Vent placement to maximize air intake and emission of hot air from a greenhouse

through the entire greenhouse. The vents can be opened in the daytime and closed at night so that plants won't be subjected to excessive temperatures. This daily practice is certainly needed for cold frames as well. The vents permit fresh air to circulate, and there should be fresh-air inflow throughout the growing season. The optimum is for relative humidity to be in the 40 to 85 percent range, and this takes careful management in much of Appalachia and other parts of the eastern United States. Higher humidity causes disease to spread and chills plants on cold winter nights.

Often greenhouses are not used in the summer even though the high summer temperatures can transform the greenhouse into an ideal food or herb dryer. If the greenhouse is used as storage space, volatile chemicals or any materials that could host pests or diseases must be kept out. Keep the extender clean and wipe the glazing from time to time. Similarly, bugs, rodents, and other varmints must be kept out. Any water containers used for thermal storage must be well sealed to prevent mosquitoes from breeding. The gangs of ladybugs that seek out indoor shelter in the fall can be gathered up and released into the greenhouse to help control aphids and other pests. The practices of integrated pest management are ideally suited for greenhouse management to reduce any potential health risks. Nontoxic, low-VOC (volatile organic compound) exterior paints are recommended for indoor use because of their extra durability; airing out the building after painting is a necessity.

Virtually every greenhouse has a sign asking visitors to keep doors closed in winter to retain heat. Sometimes greenhouse builders design adjacent air locks for this very reason. A small curtain hung near the door is also helpful to keep cold winds from sweeping indoors. Insulated window shades to minimize nighttime heat loss can also be used with greenhouses.

No vegetables or herbs grow without some tender loving care. It need not be done often, but gardeners should keep an eye on plants in the greenhouse just as they look at the garden every day. No matter how much care is taken, some plants will get diseased and will have to be either treated or removed along with any dead materials. Watering should be done once or twice a week rather than at irregular or overly frequent intervals. All diseased or dead plant material must be pruned and removed as soon as it is noted. Enough space between plants to allow for air movement is essential to keep funguses from getting started. The soils for containerized plants can be replaced with disease-free and weed-seed-free soil media. In larger greenhouse beds, topsoil can be replaced every two years with compost.

PRODUCE CHOICE

Many longtime gardeners prefer a mix of vegetables, herbs, and flowers for beauty and for natural pest control. That healthy garden practice may be extended to indoor growing.

Within the greenhouse or cold frame the experienced grower attempts to fill vacant space with living things. Potted plants are placed in the empty corners and hanging plants fill the air space above the beds. Taller plants and trellises are placed in the back area (north side) and smaller plants up front in a stair-step fashion. In selecting a combination of produce varieties, the grower considers plants that need more moisture for beds and those that need less for pots and individual containers. Some plants, such as fennel, should be omitted because neighboring plants find them poor companions.

Growers think differently when growing in the greenhouse or cold frame rather than outdoors. No one would plant heavy feeders such as corn or sunflowers or bulky crops such as potatoes or turnips in a greenhouse. Beans and peppers, which are highly sensitive to cold weather, should be avoided because cool weather will cripple them.

Summer crops that are still growing can be transplanted just before frost to a greenhouse or cold frame—a premier example of a true season extender. Move plants that can be transplanted without undue shock, such as hardier cherry tomatoes, the Appalachian "Tommy Toes," as well as celery (which can withstand temperature fluctuations better than wind chill), kale, mustard (though this is susceptible to disease later in the year), and a number of short-season greens: various types of lettuce, mizuna, sorrel, corn salad, and arugula. Endive may continue for some time as well, and spinach is a good late-autumn crop. Melons, cucumbers, peas and other legumes, and many root crops have not proven suitable for fall transplanting.

For longer-term winter and early-spring cropping, a proven producer for greenhouse growth is Swiss chard. Some growers regard it as the premier green of the greenhouse. Along with broccoli, chard can provide many highly nutritious meals throughout the limited-light season. Many gardeners find that mints, parsley, and dill add flavor in the winter months. Oregano can be left year-round, as it endures summer heat and also thrives in winter. Chives and garlic do well indoors in containers, as do basil, scallions, and radishes. For the most part, except for moving a few beets and carrots indoors for color, the grower can heavily mulch and leave bulky root crops—turnips, rutabaga, kohlrabi, salsify, Jerusalem artichokes, and horseradish—outdoors in milder eastern American microclimates. Other brassicas can be added to the greenhouse mix, including decorative kale plants.

A word of caution: recent research has found a heavier concentration of nitrates in the leaves of midwinter greens during the shortest days of winter. (Nitrates may under certain conditions be converted to nitrosamines, which are possibly carcinogenic.) One solution is to fertilize very sparingly; another is to harvest mostly when more light is available in autumn or in the very late winter and early spring. Avoid commercial nitrogen fertilizer in the greenhouse and instead use compost and diluted urine (urea) when plants need nitrogen.

Some growers cultivate certain berries in the greenhouse, especially containerized strawberries. Others add the potted semitropical cherry, which thrives even if the temperature drops below freezing for short periods. Others suggest miniature oranges and other citrus fruit. But for the most part, herbs and vegetables have served as the prime greenhouse produce.

CHAPTER 12

Wildlife Habitat Restoration

Appalachia is one of the most ecologically diverse regions of North America. This region's landscape has been drastically modified by European settlement and subsequent exploitation of natural resources such as timber and wildlife, resulting in degraded systems and extirpated species. Efforts to repatriate lost species, such as elk, fish, otter, wild turkey, and peregrine falcon have been successful and created harvestable populations in some cases. However, despite these successes, concern remains about the future of Appalachia's rich natural heritage as human population growth increases with concomitant demand for natural resources. Moreover, as resource extraction technology has been more efficient during the past few decades, its impacts have increased in spatial and decreased at temporal scales. The current challenge of conservationists is to encourage society and industry to develop cooperative solutions that both meet resource demands and protect natural systems.

Jeffery L. Larkin, conservation biologist,
Indiana University of Pennsylvania,
Indiana, Pennsylvania

Wildlife has been struggling to hold on to its habitat in many parts of America; the demise of the passenger pigeon and the destruction of the bison herds on the Great Plains apparently were not strong enough warning examples, although some groups sprang up to advocate conservation and preservation. Wildlife groups followed in the wake of conservationist movements of the early twentieth century. The Audubon Society campaigned successfully to stop killing birds for hat plumage; the Sierra Club developed tours to see national wonders; wilderness areas and national parks were created by legislation. The advent of the World Wildlife Fund and other conservation groups made large cats and other land and sea mammals, especially in Asia and Africa, a cause célèbre for a wider group of protectors on a global level. However, there are plenty of birds, mussels, and amphibians in the eastern United States that are endangered. Our fairly tough federal endangered species legislation came none too soon, but proponents have to constantly fight to keep its regulations from being watered down or eliminated.

SHARING A LIMITED HABITAT

Creating and preserving wildlife habitat can be difficult work at the practical level. No matter how one strives to design unimpeded wildlife reserves and migratory routes, the competition between human beings and wildlife for living space is fierce. The best that can be hoped for is a compromise and a procedure that permits local residents to benefit from conservation and from the tourism potential of well-managed wildlife reserves.

From whales to monarch butterflies, wildlife habitat is under attack from hunters, developers, and greedy resource extractors. On the other end of the spectrum, a few species—generally game—have multiplied until they have become worrisome pests: white-tailed deer, "wild" turkeys, and Canada geese. In fact, these three species are perhaps more numerous now than at any other time in American history. Deer can get around many fence barriers and nibble on suburban shrubs and flowers; so-called wild turkeys can lessen the reseeding rates of American ginseng in the forest understory; and geese can add unwanted manure to lawns, sidewalks, and golf greens—and even invade zoos, as at the Milwaukee Zoological Garden. They, along with other opportunistic migratory-turned-resident fowl, feed on Midwestern cornfields and settle on winter bodies of water all over the nation.

Why fly south when the fare is so lavish at home? Without some population controls, these increasing numbers of wildlife can become nuisances.

Busy families of beavers make life tough for property holders with choice natural streams. The ever-active animals chop down trees, build dam homes, and establish a beaver-order community wherever they want. John Davis, cofounder of the Wildlands Project, is willing to tolerate these animals near his humble forest abode near Westport in northern New York. He is a friend of the beavers, allowing them to work at will right outside his residence as long as they do not drop trees on his house (they missed once by inches). However, they seem to be creatures bent on establishing their own habitats—do they vaguely remind us of any other species? We might settle on a middle course between trapping and killing them and allowing them to take over. Is this one place where trapping and shipping to distant wilderness might be called for? Or what about the reintroduction of natural beaver predators such as the eastern cougar or gray wolf?

John Davis, director of conservation for the Adirondack Council, writes eloquently of the need for wildlife protection:

> Cougar, Lynx, Gray and Red Wolves, Elk, Bison, Passenger Pigeon, various fish and mussel species, and American Chestnut are among the many species that Euro-American colonists have exterminated from much or all of their ranges in the Appalachian Mountains and eastern North America. The reasons for restoration may be broadly classed in three realms: ecological, practical, moral. Lessons from all these realms tell us that we must protect much more land and water and with much greater habitat connectivity if we are to enjoy the fruits of truly sustainable natural and human communities. Ecology tells us that natural communities start to unravel when predators, pollinators, seed dispersers, excavators (woodpeckers, burrowing rodents), builders (Beaver, Muskrat, birds), or major prey species (including nut-bearing trees) are eliminated. Ecological health depends upon a full range of native species in functional, viable populations.
>
> Practical concerns include unnaturally abundant prey populations, especially deer, now generally lacking their main

> predators, becoming a menace to gardeners, spread of disease (e.g., Lyme, rabies); proliferation of biting insects (naturally held in check by birds, bats, and amphibians); and diminishment of pollination and other ecosystem services. The loss of music—bugling Elk, howling wolves, splashing trout, warbling birds—is a concern that links the practical with the ecological and moral realms.
>
> Morally, the case for inviting back the creatures we've driven away stems from our duty to honor and protect Creation. People of most faiths and ethical frameworks the world over agree in principle, if not in practice, with what we know in our hearts to be true; wild Nature, original Creation, is good and right, and deserving of protection in all its myriad forms, for its own sake, for our sake, and for the very Creative Force from which this wondrous diversity derives and to which it points. Bring back the big cats and wolves and bears and buffalo and salmon. . . . Let's go wild!

Often the wildlife balance is disturbed, and some species may proliferate beyond the carrying capacity of the particular ecosystem, creating a situation that is problematic for both the species and its habitat. If the niches of the carnivores are vacant (and we lack the foxes, packs of wolves, and wildcats of old), then the herbivores can sometimes reproduce beyond their own best long-term survival interests. Havahart traps, made to capture wildlife in a humane manner, may not be appreciated by those living near or on the property where the nuisance wildlife is eventually released. A far more humane alternative is to reintroduce the carnivores that will help return nature to its original balance. We must learn to welcome back the wildcats and red foxes, which control rabbits and other small mammals, and black snakes, which can control mice. That welcome needs to extend to black bears and to the mountain lions or lynx or pumas, and that is best achieved by ensuring the restoration of their habitat.

Finding a balance between the interests of wildlife and humans is not always easy. For example, gardeners living near wildlife habitats realize that some of their vegetables are choice delicacies for everything from rabbits and groundhogs to raccoons and turkeys. They find that having a good fence is not sufficient. A watchdog is better, but a dog can be quite aggressive and may not be able to guard distant gardens. Living with wildlife takes some creativity and yet one soon finds that some varieties of vegetables (especially spicy ones) are not palatable to most wildlife. We have found that ringing patches with beds of spicy mustard will dissuade the rabbits, which gravitate to beans and other delicacies. Beware of wildlife favorites such as corn. Some gardeners simply grow enough to satisfy animals such as raccoons, but it is hard to grow enough for hungry groundhogs. Deer are finicky but generally leave melons and squash alone—although bucks have been known to ram pumpkins with their antlers to break them open and get the seeds.

Deer avoid double fences erected about 6 feet apart. Few animals like onions, garlic, the nightshades (potatoes, tomatoes, peppers, and eggplant), some of the brassicas (cabbage, collards, cauliflower, and kohlrabi), turnips, radishes, okra, or most squash. Most animals fancy beans, with carrots and lettuce as second choices. Keep these vegetables in beds near the watchdogs.

As a last resort, if you like the pioneer dish burgoo (a kettle of various varmints cooked for 24 hours with potatoes and onions), you'll find garden raiders make excellent local food for meat eaters bent on eating local produce, even if it is a notch up on the food chain.

MINIMIZING HABITAT DISTURBANCES

Living with wildlife is a blessing and joy and also an art. We learn to see animals as friends and realize the heavy stress that many of these species suffer due to human development and fragmentation of the wilderness.

Restoration is more than planting choice wildlife feed or building feeders and nesting areas. The opposite of construction is calling a halt to building—roads, fire lanes, wilderness camping areas, scattered housing developments. Throughout the twentieth century, roads were carved through pristine wilderness areas, threatening wildlife nesting and migration routes and creating corridors where invasive plants are encouraged. And traffic leads to animal roadkills, poaching, and greater habitat destruction. Furthermore, roads in the wilderness attract scary aliens such as off-road vehicles and snowmobiles and their careless drivers.

I once performed an environmental resource assessment near a medium-sized eastern seaboard city. The directors of the property (a convent) were very proud of the greenery of their urban 20 acres, a fragment of the original forest. Upon entering the fragmented area, however, I found that it contained a host of invasive exotic species common in the eastern United States and that the green vegetation was choking out the oaks and maples that had been part of the forest before the fragmentation. Urban greenery fooled the property holders, who did not realize that such undeveloped islands of greenery are already damaged from assaults on all sides. Most wildlife needs large tracts of undisturbed land, which are becoming ever more scarce today.

Nature trails can be further sources of fragmentation unless conscious design makes them less infringing. These trails should avoid fragile areas such as old-growth forests, rock moss communities, and many wetlands. These pathways into the wilderness offer opportunities for hiking, observing, experiencing, and photographing wildlife, but their use may need to be restricted. Habitats of sensitive species should be undisturbed.

A proper trail is designed to deliver a complete nature experience, well built to minimize erosion, well equipped with bridges or steps where needed, and well described so that the participant will understand all of the features worth seeing or feeling (for blind hikers). Nature trails may be classed and publicized according to the degree of exertion they require, and portions may be reserved for people who are less mobile. The trail surface may be hardened with traditional paving or with new plastic substances that mix with soil to create a firm, all-season base even able to accommodate hikers who use wheelchairs or walkers.

Markers, signs, and permanent or mobile audiotape units prove serviceable when there are no human guides.

Unfortunately, roads and trails are also sometimes inviting to those who can damage the wilderness. Motorized vehicles (snowmobiles, dune buggies, and other all-terrain vehicles [ATVs]) often travel over posted lands with little regulation or traverse federal or state wilderness areas with inadequate enforcement. In response, those responsible for trail maintenance must do more than post areas. They may have to construct barriers such as steel gates with steel posts set in concrete, fencing, huge boulders that cannot be pulled out with trucks or tractors, tank traps, or "Texas crossings" (zigzag fencing that allows hikers to pass easily but not horses or ATVs). Overuse of trails may frighten away certain nesting birds such as the cerulean warbler in the Appalachian mountain range.

Most trails need tender loving care to keep them in good shape for hiking. Trees fall, signs are damaged, leaves cover the walking surface, and erosion may occur. A major threat to nature trails in times of reduced budgets is lack of supervision, maintenance, and enforcement. Here volunteers can make a difference—and wilderness work is inviting, for it is good outdoor exercise, requires teamwork but only moderate skills, and the result is an artifact that has ongoing value for others..

Nature experiences such as those one can experience along the Appalachian Trail (see figure 12.1) are good for the human adventurer, but some wildernesses should be off-limits to even the most careful. The pressure to open more and more lands to human intrusion is increasing, but can't virtual experiences (books, nature observation stands, films, wide-screen viewing) suffice in many circumstances? Visitors with the best of intentions can exceed the human carrying capacity of a wilderness. Banning fires, requiring that visitors carry out all the litter they generate, forbidding wildflower gathering or feeding wildlife, and discouraging overland (off-path) hiking are all measures that may need to be considered to reduce damage to the wilderness. More-intense experiences with naturalist guides in smaller desig-

Fig. 12.1. A hike along the Appalachian Trail affords the visitor views such as this.

nated areas can be almost as valuable as wilderness experiences, for committed naturalists can be excellent interpretive guides and role models.

CREATING HABITATS

Because biodiversity adds to total environmental health, the goal for concerned residents is to encourage a healthy balance of the natural community and a wide variety of wildlife. This biodiversity must be reestablished through habitat reclamation at every level, from the individual backyard to the international level. Local or regional nature centers, reserves, and parks can promote biodiversity and provide environmental education. State and national parks play a significant role through educational displays and reminders of the fragility of natural surroundings. Such groups as "Friends of [name your favorite park]" provide valuable assistance to restore wildlife habitat and reclaim wilderness.

Leaving uncut tree clusters or wildscape at fencerows or in backyards encourages the reintroduction of wildlife. Certain plants available in specialty stores attract birds and other wildlife. Feeding birds has been one way people encourage a comeback; organic lawn-care techniques attract robins and other birds quite quickly. Phil and Mary Stern of Allenspark, Colorado, have an active backyard feeding program and attract twenty types of birds and squirrels and coyotes in winter. Some backyard wildlife promoters, to the utter annoyance of neighbors, install salt blocks for deer.

Everett Reems, a hardy mountaineer in Randall Cove (a tributary of Bald Creek in the Big Sandy Mush Creek Community in the Newfound Mountains of western North Carolina), has been feeding wild deer at his house for several years. Ironically, his living room and kitchen walls are lined with mounted deer heads, as he is a well-respected taxidermist. But the wild deer don't seem to mind. However, we must caution that making wildlife dependent on handouts is as bad as practicing human charity without restrictions.

Protecting high-quality habitat may prove more effective than enabling dysfunctional animal behaviors. For instance, too many chickens may decimate earthworm populations used for composting; untrained and/or unrestrained domestic or feral dogs may cripple or drive away wildlife; domestic or feral cats can play havoc with attracted bird populations.

It is also possible for an individual to create a bird sanctuary. A bird sanctuary is land declared to be such by the property holder(s). One need not wait for an official organization to formally authorize the designation. Bird sanctuaries allow birds to nest, rest, and feed without risking harm from humans or their pets. Feeding birds is frowned upon by some purists, but it may be better regarded as a countermeasure to the vast destruction of natural habitat in urbanizing America. Hummingbirds can be attracted by sugar-water feeders, and butterflies favor certain flowers. Providing nesting places can complement bird feeding and watering, but in overly fragmented areas the birds are vulnerable to cowbirds, which lay eggs in others' nests, leading to competition among babies and excess labor on the part of the nests' owners. Very small sanctuaries are sometimes only of limited usefulness, but they often are some of the only resting places left to neomigratory tropical songbirds.

Long Branch has a "Wildlife Habitat" sign (from Defenders of Wildlife) stating that the center is a place where wild flora and fauna are respected. In these 1,600 acres of forested western North Carolina mountains can be found a host of animals: black bear, wildcats, squirrels, rabbits, raccoons, skunks, ruffed grouse, various snakes, trout, and a host of migratory birds. But wildlife is challenged by the proximity to expanding Asheville, where developers compete for potential pristine building sites. Other threats come from invasive species of plants and from the equally invasive exotic all-terrain vehicles, which destroy the tranquillity and health of wildlife habitats.

The Mary E. Fritsch Nature Center near Livingston, Kentucky, strives to demonstrate native wildlife treasures to Appalachian residents who regard these forested lands as exploitable. The center contains a demonstration building and 3 miles of marked nature trails. The staff has labeled over one hundred trees and woody species all naturally growing in this mixed mesophytic forest. Likewise, about forty wildflowers are identified. The coves, Rockcastle River valley, and surrounding hills, including the adjacent Daniel Boone National Forest, offer habitat for a variety of species. Educational programs teach youth about local wildlife. The center provides information on over two hundred species of birds, both migratory and permanent residents.

Wilderness areas are found in both sparsely and heavily populated parts of America, including Appalachia, and need to be protected and expanded. Platforms and viewing towers could allow visitors to observe fragile areas without entering them and damaging the fragile ecosystem. Thus through binoculars, platform signs, audiotapes, and handouts the visitor can have a satisfying experience without trampling the highly diverse understory plants and fragile ecosystem.

Large undisturbed or reclaimed areas could be used in the reintroduction of threatened wildlife. A recent attempt to restore red wolves *(Canis rufus)* in the Great Smoky Mountains National Park (GSMNP) failed, in part because visitors' pet dogs transmitted a fatal parvovirus to the wolf cubs. However, the efforts to reintroduce the red wolf have been highly successful in the 152,000-acre wilderness of the Alligator Wildlife Refuge in eastern North Carolina. The reintroduction of the American elk *(Cervus elaphus)* into Appalachia continues to be a success story, especially in Kentucky, where over five thousand elk now roam, These elk enjoy a 90 percent breeding success rate and a 92 percent calf survival rate. Elk are also being released in the Cataloochee Valley area of the GSMNP.

A total of 267 river otters *(Lutra canadensis)* were relocated to eleven rivers in western North Carolina by 1995. The success of the project has been monitored by stream surveys. Based on sightings of animals, the project seems to be highly successful, with river otter populations expanding and reproducing.

The arboreal weasel, the wejack or fisher *(Martes pennanti)*, is enjoying a successful reintroduction into Appalachia from northern Wisconsin. Fishers are famous as voracious predators and are an excellent biological control on egg-eating animals like skunks, raccoons, and possums, varmints that prey on ground-nesting birds such as ruffed grouse. Over twenty breeding pairs of fishers were released in the 80,000-acre Catoosa Wildlife Management Area of eastern Tennessee. Fishers disappeared from that part of Tennessee woodlands in the nineteenth century following intense trapping and logging. The reintroduced fishers are thriving, and plans are under way for similar projects in the Cherokee National Forest and the Royal Blue Wildlife Management Area in Tennessee.

Yes, the buffalo (American bison) roam. These majestic animals of the Great Plains were travelers that crossed the Ohio River to obtain salt at the various licks (saltwater springs) in Kentucky's hills. Reintroduction is somewhat difficult even where land is plentiful because of the bison's propensity to travel about and move through fenced areas when they so desire. Finally, efforts are being made to protect the rare mountain lion or eastern cougar where sightings are made.

CHAPTER 13

Nontimber Forest Products

> *Witch hazel is not only a most interesting shrub in itself, but also has connected with it many legends. From its forked twigs were made the divining rods by which hidden springs of water or mines of precious metals were found. It was firmly believed that the twigs would turn in the hand when the one who held it passed over the spring or mine. More recently, its fresh leaves and twigs were used in large quantities for the distilling of the healing extract so much in demand as a remedy for cuts and bruises and for chapped or sunburned skins. It is said that the American Indians first taught the European settlers concerning its medicinal qualities.*
>
> Gene Wilhelm, retired professor,
> Slippery Rock University,
> Slippery Rock, Pennsylvania

Applying appropriate technological concepts to Appalachian forests involves the choice of what is to be harvested or removed (this chapter) and actual harvesting practices (see chapters 14 and 15). Let's review the types of nontimber forest products (NTFP) and discuss decision-making processes related to their gathering and use.

TOURISM

A recent scientific publication on NTFP neglects even to mention the most important Appalachian forest product: the experience that visitors enjoy when using the scenic forest for recreation. Forests can be observed by the visitor, camped and hiked in, climbed over, traversed by bike, water tube, boat, or motorized vehicle, or just enjoyed through videotapes or photographs. Though the U.S. Forest Service acknowledges that recreational experiences are the number one NTFP, recreation receives little attention from NTFP experts because it is not a commercially quantifiable commodity. Yet the 11 million annual long- and short-term visitors to Appalachia represent immense potential economic value. Tourists spend money for services so that they can enjoy the forests in widely varying ways. And these tourists are loyal; the great majority return year after year because they enjoy the scenery.

NATIVE FOODS

From before the beginning of civilization, hunter-gatherers have stalked the woods searching for food to keep them and their families alive. Gathering practices continue today, especially at the edge of or outside the traditional economy. Native foods from forests have economic value based on commercial sales, but these are small compared to the noncommercial supplemental value these foods have for many residents and visitors. Some general food classes include:

greens (see chapter 15)

fruits (e.g., crab apples, persimmons, papaws, mayapples, fox and other wild grapes, wild cherries, plums)

nuts (e.g., black walnuts, hazelnuts, various types of hickory nuts, butternuts, chestnuts, acorns)

berries (e.g., blackberries, raspberries, wild strawberries, blueberries, elderberries, dewberries) mushrooms, both native varieties and imports such as shiitakes, which have high commercial value and can be grown on oak, black locust, sweet gum, and red maple logs

honey for use as food and to make beeswax

maple and other tree-sap syrups

roots and barks (sassafras and others)

wild game and fish

FUEL

Firewood is a major NTFP even though it comes directly from the portion of the tree that yields the timber. Firewood has great commercial value, especially in forests near large urban and suburban areas. Chips, brush, and other tree wastes are used as fuel in certain wood-burning devices, but coal, natural gas, and oil are not products of the forest except in the most indirect manner. Massive resource-extraction practices can harm the forest, especially the surface mining of coal or overharvesting of timber.

MEDICINALS

Plants like American ginseng, goldenseal, yellow root, and bloodroot (see chapter 15) have become valuable commercial resources from Appalachian forests. Poaching of such products is quite common today and needs to be controlled to keep the plants from being threatened or endangered. Unsustainable extractive processes are the result of the high price of ginseng root and the growing market for other medicinal herbs.

HANDCRAFT AND HOME PRODUCTS

The range of handcraft products is quite wide, limited only by the imagination of the crafters. Grapevines and other nonpoisonous vines, even kudzu, can be woven into baskets and Christmas wreaths. Wood knots and burls can be carved into bowls and statues. American bamboo or "river cane" (the only native bamboo species) can be threaded into mats, racks, and furnishings (see figure 13.1). Grasses and dried flowers can be made into bouquets for commercial sales or home decoration. Tree bark has a variety of uses, from flavoring to dyes.

SEEDS AND PLANTS

Picking flowers, seeds, or plants may seem harmless, but the practice can do considerable damage when the species are threatened or endangered and the practice widespread and frequent. In

Fig. 13.1. Stand of American bamboo (river cane) at ASPI

some densely populated nations no gathering of any type of forest product is allowed without an official permit. However, in the fragmented and urban forest, one finds many exotic and invasive species, whose removal is essential to preserving the well-being of native species. Digging out wild chicory for a dried-root coffee substitute would be a benefit to the biodiversity of the region. The same goes for harvesting garlic mustard for sale and use. And how about harvesting tons of kudzu so the foliage can be used for livestock fodder, the vines for craft articles, the roots for cooking, and the blooms for preserves? There are at least 7 million acres of kudzu in the South, including large parts of southern Appalachia, as of this writing.

CHIPS AND OTHER NONTIMBER TREE PRODUCTS

Removal of small trees and brush to make wood chips for export or wood pulp for paper could result in severe environmental damage. The heavy machinery used in such removal compacts the understory, and the near total extraction of much of the vegetative materials depletes the soil of precious organic matter. Chips, which are made into composite board and other building materials, are becoming an important commercial NTFP. Waste materials from the timber-making process such as sawdust (used in compost toilets), slabs (used for siding, fencing, and other construction), and slash (used for erosion control) could also be considered NTFP.

A NONTIMBER FOREST PRODUCT ETHIC

Grandma's adage "Take a mess, don't make a mess" is worth some reflection. Have a taste or a dish of spring greens, but don't tear up the land and rip off an entire patch of greens in doing so. Be slow at taking too much of a good thing, especially in trying to commercialize any particular NTFP. Consider another aphorism: "Take only what you need and leave the rest."

Developing an ethic of nature experiences could begin by asking: Should any recreational activity be allowed in the forest? What about encouraging the hunting of the increasing numbers of wild turkeys, which can damage the forest understory? How about the use of off-road vehicles? What about metal spikes left on rock-climbing surfaces? How about indiscriminate cutting of firewood on camping expeditions? Or should we even enter and impact wilderness areas at all? Should we feed the wildlife and make the animals dependent on handouts?

Nature experiences vary in impact depending on the way they are practiced rather than on any hypothetical forest "carrying capacity." The off-road vehicle operated by someone disconnected from nature is potentially dangerous. One such off-road vehicle carelessly operated in muddy conditions could do far more damage than a hundred prudent hikers or a thousand sightseers. The philosophy of appropriate technology calls for restrictions on damage to trees, the understory, and wildlife, as well as improvements to the safety of recreationists.

Economics is an important part of the ethical equation, as when a cash-starved Appalachian woodland owner sells a long-held inheritance to a fly-by-night chip mill operator. It's a common practice these days in the Deep South and is moving north. What is often sold is the forest commons, meant to be enjoyed and used to improve air quality for all the people, not just the landholder. Unfortunately, the forests are so valuable both for timber and for commercially salable NTFP that virtually any portion of a wild landscape is at risk, especially in areas where pine plantations are now in vogue. Do the sellers reflect that the chain saw cuts far more than yesteryear's crosscut saw did and that heavy machinery makes a far greater impact through compaction of the understory than the horse? Supplemental incomes from a variety of economic sources under free-market conditions are still possible, but greed must be controlled.

For example, St. John's wort is used by many to relieve mild depression, but many potential gatherers don't know what to look for, where it grows, when to harvest it, or how to pick it, much less how to preserve it. Giving unscrupulous novice harvesters a small amount of information is dangerous, as is connecting these potential poachers with unscrupulous marketers. Perhaps a half million central Appalachians own tracts of woodland, and some owners are tempted to sell their forest products to chip mills, producers of medicinal supplements, or marketing warehouses. They may not realize the dangers involved.

One approach is to offer educational programs on the valuable uses and the dangers of overuse of NTFP in Appalachian forests. This could lead to knowledge of the forest for its own sake. In our visits to the woods we grow in appreciation of the diversity and complexity of interacting systems and begin to want to preserve them. We learn to champion our marvelous forest treasures and help others acquire deeper respect for fragile forest systems.

Such programs could teach that some forest benefits are to be shared by all: scenic views, climate moderation, soil retention, flood prevention, and enhancement of the water table. Other benefits are more direct, such as land use for environmentally sound recreational purposes and the gathering and consuming of NTFP. Information on the commercial feasibility of NTFP is now provided by colleges, technical schools, extension services, and for-profit groups. These programs must always be coupled with safeguards for preserving the forest ecosystem.

Programs may also pass along the information that the presence of specific NTFPs can add considerable value to land, as one ginseng grower has found. But a smorgasbord approach to disseminating NTFP information is fraught with environmental problems such as the endangerment of fragile plants. Incomplete information can lead to improper harvesting practices. A few years back, the late expert Cherokee naturalist Hawk Little John denied a request to document his gathering of wild herbs because of the likelihood that the information would fall into the hands of potential poachers.

It is also dangerous to overemphasize forests' economic potential. Conceding that the presence of NTFP increases land value implies that economics should take precedence over ecological considerations. Land has great value with or without a particular plant, and it should not be defined solely or principally by commercial considerations.

Additional research is needed to identify which NTFP are appropriate for development. A general, superficial approach to NTFP could be more fundable than more narrowly focused research. The gathering and dissemination of general NTFP information appears "cool" to some funders, while researching single products is hard work, takes precise observational skills, and may produce negative results. Appropriate research should balance economic and ecological considerations and make a distinction between products that should be encouraged for harvesting, such as invasive species like kudzu; those for which harvesting would be tolerated, like species (blackberries and hickory nuts are examples) that can continue to flourish even with some harvesting; those for which harvesting should be regulated (ginseng); and those that should be protected from harvesting.

How should specific target products be endorsed? A sustainable harvest pattern needs to be established, based not on greed or market demand but on the ability of forests to regenerate the product in a sustainable manner. Primitive peoples from the Eskimos to the Appalachian Native Americans used harvest practices that allowed the products to flourish for future generations. The ethic of sustainable use of natural resources must be shared at every level of our culture.

One regulatory safeguard is to have approved NTFP marketed through the use of a marketing card that indicates that the product was grown on one's own landholding or on property where the owner permits the harvesting. This combats the very dangerous practices of poaching and black-marketeering. Tobacco, an unhealthy product, was marketed through controlled programs for over sixty years. Herbs and many other NTFP could be sold in a similar manner with a high degree of success. NTFP research should be focused on how to obtain a low-impact, high-quality, marketable commodity that is virtually wild when cultivated or encouraged within the forest ecosystem. Woodland owners, residents, and lovers of forests need to unite and protect all threatened NTFP from possible poaching or unsustainable harvesting practices through a viable marketing system controlled by state and/or federal governments.

A PLEDGE TO SUSTAIN FOREST ECOSYSTEMS

Everyone can help to sustain forest ecosystems by taking the following pledge:

> *I pledge to be a champion of a healthy, diverse forest and resolve to know plants that comprise the forest understory.*
>
> *I affirm that forests and their wildlife are the commons of all people. We are stewards of all lands under our temporary control and not absolute arbiters of particular forest practices. We are accountable to the laws and regulations that preserve forests for future generations.*
>
> *I believe that plants, wildlife, and other NTFP have an intrinsic value apart from current or future commercial use. Their presence and biodiversity are worth espousing and preserving.*
>
> *I respect the private ownership habits and boundaries of other woodland property holders and will not trespass or remove plants and wildlife from these lands without explicit permission.*
>
> *I will participate only in nature experiences that enhance the forest and are safe for the users, and I will work to eliminate improper or excessive recreational practices.*
>
> *I will refrain from telling others about specific species that could have economic value unless I am certain the person will not harvest or encourage others to harvest the plant in an unsustainable manner. I recognize the propensity of many to be imprudent in their relationship to NTFP.*

I will not introduce exotic species into the forest that may threaten the native ecosystem, nor introduce species that will affect the genetic pool adversely, nor will I allow to go unchallenged the introduction of such species by local marketers and others.

I will not engage in or allow excessive or unsustainable timber harvesting on lands under my control.

I will not remove or encourage others to remove plants or animals that are rare, endangered, or threatened and will report such practices to proper authorities when observed.

If physically able, I pledge to improve the quality of my own or neighboring woodlands by careful forest management that includes removing exotic species that threaten the biodiversity of my forest understory.

I pledge to encourage the proper harvesting of economically viable NTFP that meet all the criteria for sustainable harvesting and continuation of forest health. Personally, I will only harvest from the forest limited amounts of NTFP for occasional enjoyment or for modest economic gain and do so in a sustainable manner.

CHAPTER 14

Silvicultural Practices

Few issues are more critical, or more hotly debated, than proper forest management and timber harvesting techniques. The long-term health of forests, streams, and wildlife, as well as personal and community economies, is at stake. On a personal level, a landowner may fear excessive damage to their woods by heavy equipment, sloppy logging, over-harvesting or undervaluation of their timber. These often valid concerns coupled with the fact that Kentucky has some of the most diverse, productive hardwood forests in the world, led my wife Beth and I to start Woodland Farms Modern Horselogging. As both landowners and loggers, we developed a list of common interests to satisfy our concerns for the forest's health on the one hand, and the need to make a living on the other.

Gary Anderson, horselogger,
Sustainable Logging and Lumber Production

Trees are a critical resource, and any action that diminishes their health and well-being is an attack on the vitality of Earth itself. Trees may appear robust on an individual basis but be part of a forested community that suffers from neglect; unsustainable harvesting practices; fragmentation through highways, development, and logging roads; introduction of invasive exotic plant species and insect pests; and private clearing.

Anglo-Saxon land-use practices have influenced our forest-related attitudes. In old England, a forest was residual land too poor to plow and cultivate. Arable lands were highly esteemed, whereas forests were regarded as wastelands. English colonists brought with them a bias against forested areas and saw fertile forested land as a challenge or as an enemy to be conquered with an ax. Good as well as marginal farmlands were cleared from New England to Florida and points west, with nonarable cleared land used for grazing. A haze is said to have hung over the southeastern United States from about 1810 to 1830 while the land was being cut, cleared, and burned over for growing cotton and other crops.

ASPI sponsored a forest commons conference in 1994 that brought together a number of experts who addressed forest practices and the need to see forests not as being owned and managed in a laissez-faire manner by landholders but as the heritage of all people. The people collectively thrive upon the forests and should uphold them as a commons like the air we breathe. This concept is more universal than conservative private-property-rights-oriented Americans are often willing to admit and is critical to any discussion of long-term forest preservation.

I come from a family of foresters who lived in villages in the forested border area of northern Alsace and the German Pfalz. When I visited this ancestral area I observed the extent to which trees have been part of my family heritage. In northeastern France and nearby Germany, forests have been well managed for centuries, though, as in other parts of the world, their current health is threatened by air pollution and other adverse conditions.

The Cradle of Forestry is a forest center operated by the U.S. Forest Service in western North Carolina. The center honors C. A. Schenck, who founded the Biltmore Forest School and taught there from 1898 to 1913. He was invited by the wealthy Vanderbilt family to bring the silvicultural experience he acquired in Germany to the United States. A conservationist philosophy had arisen during Theodore Roosevelt's presidency; the nation's insatiable appetite for forest timber was pitted against a growing realization that forest resources are limited and could be exhausted should current uncontrolled practices continue. Schenck respected nature, the manner in which trees grow, the need for tree care and protection, proper selection and cutting practices—in short, good, scientific forest management. His goal was a healthy forest that sustainably supplied timber to meet reasonable needs.

TREE SELECTION FOR PLANTING AND HARVESTING

Should one select thriving native species such as tulip poplar *(Liriodendron tulipifera)* and pin oak (*Quercus palustris*), or threatened species such as red oak, Fraser fir, and eastern hemlock (*Tsuga canadensis)*? Before the type of native species is addressed one must know the specific site and what trees grow best there, and the purpose for which trees are being grown. The site may be a cove where hemlocks have thrived for centuries or a mountaintop on which the Fraser fir has grown well in the past.

Another consideration is current threats to specific trees. The eastern hemlock and Carolina hemlock *(Tsuga caroliniana)* are under attack by the hemlock woolly adelgid. Other Appalachian trees are affected by the southern pine beetle and exotic insect species, including the European and Asiatic gypsy moth, the balsam woolly adelgid, and the Asiatic oak weevil. Current oak decline is expected to lead to a reduction in number and variety, especially in the red oak group, but not an elimination of the species. Dutch elm disease spreads slowly throughout the region and kills the relatively isolated native American elm *(Ulmus americana)*. For a complete treatment of native and exotic pests, see Southern Appalachian Man and the Biosphere's *Terrestrial Technical Report* (see Resources).

Besides the specific site and availability of trees to be planted, a planned selection addresses the question of what the tree will be needed for in two, five, or ten decades. Healthy forests may be required for visitors, wildlife, reasonable timber harvesting, or a combination of these. Long Branch recently worked with the National Fish and Wildlife Foundation to plant a number of species, including red oak *(Quercus rubra)* for acorns, black walnut *(Juglans nigra)* and mockernut hickory *(Carya tomentosa)* for nuts, and papaw *(Asimina triloba)* and persimmon (*Diospyros* spp.) for wildlife habitat restoration.

Forests are managed for beauty, recreation, water quality, wildlife habitat, and—when harvestable—timber. A black walnut may furnish nuts for years and, when managed well, be cut for veneer worth $1,000 or more per log to furniture makers. Enrichment plantings mean adding rare or valuable trees (e.g., American chestnuts) to a forest patch with the proper sunlight to allow them to flourish again in their native habitat.

To preserve healthy ecosystems, good forest management takes into consideration tree selection and all phases of tree growth and protection. (For a good discussion of forestry practices, see Paul Kalisz, *The Practice of Ecoforestry.*) Property holders can learn the condition of their forests by accompanying professional foresters on inspections of their land. Winter is a good season for such a forest education, which should include a topographic quadrangle map covering the tract. Enlarging that map allows the establishment of management units that are based on bedrock land-use history and topography. The tree stand in a management unit is estimated in acreage and inventoried for tree harvesting through accepted methods.

An inventory is performed best by use of an inventory stick, a handy measuring device that helps in forest description. Ecosystem inventories include ecologically important features such as caves, cliffs, and cavity trees, which function as dens for wildlife. Snags, logs, and treefall mounds and pits are all part of the wooded landscape. An inventory facilitates an assessment of the overall health of the forest, the types and growth of the trees, where reforestation can occur, and what species should be planted. Selection then becomes an exercise in both what to plant and what to harvest. The inventory permits an estimate of the sawtimber available in areas where sustainable harvesting can occur without harming the forest. From the results, a proper ecological restoration program can be launched.

REPLACEMENT PLANTING

Planting a single tree can be a symbolic act, showing respect for nature and one's mortality, as one's own death most often occurs before the planted tree reaches maturity. We plant for the future, whether in a backyard or in the woods. It's appropriate that such a tree planting be accompanied by special ceremonies with songs, poems, prayers, and maybe a dance or communal filling of the hole with soil. The planting may occur on a state's Arbor Day or in connection with another observance, properly publicized and documented, or at the time of an individual's anniversary.

Sizable tracts of land are being reforested in a number of states in the East because marginal farm and pasture land has been removed from production (see chapter 17). In such instances, the actual planting of the trees amid invasive exotic woody species may prove an empty gesture, especially in urban areas. Time and effort are needed to prepare land for the planting. The social and political ramifications may also be important; forests improve the value of the community, and enlisting civic, educational, and religious organizations for community reclamation projects can prove as beneficial as the federal government's programs to clean up waterways.

Tree planting takes more skill and attention than merely sticking seedlings in the ground. The seedlings should be kept moist and the roots protected from direct sunlight. A hole of sufficient size should be dug either by hand or by a mechanical tractor and plow or posthole digger. The roots need to be spread evenly about the hole to maximize the root area. Some trees require loose soil with high organic matter, and this soil must be firmly pressed around the seedling's roots with the surface of the earth at the point at which the root begins or slightly above it to compensate for settling. The stem should be left free of soil at planting and during the tree's growth.

Elementary care is necessary after the planting. Failure to provide such care can doom a reforesting effort. In Haiti, for example, unprotected seedlings planted in reforestation attempts were eaten by goats. In fact, the border between Haiti and the Dominican Republic can be defined by the lack of trees on the Haitian side and an abundance of them on the Dominican side. To give the newly planted tree a chance, one can use a plastic cylinder to guard the stem from varmints. In dry weather the planted tree should be amply watered. Trees could be given an elementary pruning to remove stems that will damage the main trunk or channel energy for major tree growth. During the first few years, the planting should be inspected to see whether the tree pit is properly filled and that vines or weeds are not choking out the sapling. There may also be a need for the added protection of fencing or other barriers to limit damage by deer and other wildlife.

THINNING AND MANAGEMENT

Few issues are more hotly debated than proper forest management, for the health of the total environment is at stake. After recent major wildfires in the American West, controversial federal legislation was enacted to remove tinder through controlled burns or extra cutting. But the need to remove exotic and invasive trees such as the tree of heaven and bush honeysuckle is not controversial. Invasive species must be discovered and aggressive eradication programs undertaken. An effort must be made to keep out poachers. So often in our environmental resource assessments, we find that people neglect proper forest management because of the pressure of other administrative duties. In some states poaching is inadvertently encouraged because fines generally are only a fraction of the value of the poached timber. I once observed evidence of recent logging on some land being assessed, and the staff showed surprise. From the size and quality of the stumps, the poachers got rich without the owners even knowing that the trees had been cut and removed.

Gardeners know that quick-growing plants can smother or choke out others. Sunlight is necessary for plant growth. Here, horticulture and silviculture are similar. Trees grow and need care. If a hickory sapling is caught in a locust or box elder thicket, the manager should determine the growing space needed for the healthy nut producer and release it. This clearing or thinning is usually done for trees grown for timber, but it could also involve releasing a fruit or nut tree for wildlife or personal use. As with thinning practices, pruning goes beyond orcharding and extends to the entire forest. Many future timber trees could use occasional pruning.

No issue is more contentious in forest management circles than allowing the accumulation of thatch, which can fuel an intentional fire or one started by a lightning strike, resulting in severe damage to life or property. American forestlands have been burnt repeatedly in the past, both intentionally and accidentally. Although forest floors do not have to be kept as clean as lawns, buildup of dry thatch can prove to be a fire risk in certain situations. Yet recently burnt-over forest areas have more bounce-back than most people imagine. State and county foresters can be consulted to determine if prescribed burns can improve growing conditions and reduce overall wildfire risk. With extra effort, overabundant supplies of dead materials in forests can be cut, chipped, and returned as organic matter to nourish the soil.

Park rangers in forested areas spend much time fighting invasive species such as kudzu, Oriental bittersweet, privet, and multiflora rose, which can choke out mature trees. This management work is required for both urban and rural forests. After habitat destruction, some environmentalists regard invasive exotics as the most serious environmental threat to the planet's ecosystems. Some regard the fox grape and other wild grapevines, although native, as damaging to certain trees and include these vines in their management plan, but that is a questionable practice. The measurement tools used depend on the invading species, but herbicides should be used in highly targeted situations and only as a last resort.

As for protecting forests from unwelcome activities, that is difficult in this age of off-road vehicles. The best protection for a forest may be residents living in or near the forested area who are willing to report trespassers to authorities.

HARVESTING

If we want to continue to use lumber as a building material, trees will have to be cut. But trees are a renewable resource. In the Menominee Indian Reservation in northern Wisconsin are verdant, seemingly untouched forestlands where a quarter of a billion board feet of lumber has been removed over the past century. Forests can sustainably yield both nontimber forest products *and* timber. It takes a long-term management team that respects, loves, and controls its forests. Many modern forestland holders feel the pressure to sell their trees to lumbering operations. The price may be tempting, but there are many questions that should be asked first. Has the timber been deliberately undervalued? What about the environmental record of the logging operation? Is there a sensitivity to making use of dead, damaged, or otherwise low-quality timber first? Will the logger fell timber with as little damage to surrounding trees as possible? Will the skidding technology employed reach scattered trees throughout the woods while doing as little damage as possible? Will the remaining trees be damaged? Will the understory remain in a healthy condition?

Forests can be harvested profitably and sustainably if some basic ecoforestry rules are followed. Management plans should be based on uneven-aged harvesting through single-tree selection. Trees should be felled directionally. Sawlogs should be skidded out of the woods by draft horses whenever possible, using a logging arch, a special tool that helps lift the log and keeps it from gouging the earth during removal (see figure 14.1). Logs should be sawed on-site for grade, or highest quality, using an efficient portable

Fig. 14.1. Horselogging in Kentucky

band saw. Lumber should be graded as it is cut and sawed. Lumber should be kiln-dried, preferably with a solar kiln that uses renewable energy. Slab and trim wood should be sold for nontimber uses. By using these methods, the landowner gets twice the money per tree and the logger/sawyer gets five times the money per tree. The value of the log is increased two to four times, the forest is not reduced to a monoculture or tree farm, and the timber can be cut on a fifteen- to twenty-year rotation, with each harvest increasing the value of the forest. Furthermore, the viewscape is unimpaired and the wildlife habitat is sustained.

The dangers involved in tree harvesting should be understood. Many of us have had a direct experience of near misses or accidents in tree-cutting operations. Prospective loggers should take a master logger's program such as that offered by Gary Anderson of Constantine, Kentucky. Other programs can be found by contacting state foresters. Felling trees is dangerous and, for the inexperienced, so are working horses and running the sawmill. Proper organization will make the difference between a living wage and a loss. It takes a combination of factors to harvest trees well and to the benefit of all: good-quality lumber, cooperating weather, a strong work ethic, proper horsemanship, an inviting terrain, and, if the timber is not to be used by the parties doing the operation, marketing skills.

Forest work starts with planning and planting and gets progressively more involved with management and harvesting. One cannot expect an amateur to be skilled at all these operations, but all can learn to become good managers of the forests entrusted to them.

CHAPTER 15

Wildcrafting

As a longtime sanger [ginseng harvester], let me say that it's fun to dig ginseng, but it takes some work and care. Many folks come out when we're not looking and they steal roots. If there's too many of these Saturday night diggers, there won't be much ginseng left. God gives us enough to make a living, but we've got to do our part to keep it growing for our grandchildren. That's the way I see it.

Don Surrett, North Carolina wildcrafter

Wildcrafting is the practice of gathering or extracting specific native plants or parts of those plants (seeds, stems, leaves, roots) for practical purposes—processing, selling, or utilizing them for one's livelihood. This is a general definition; the term is not found in *Webster's New World Dictionary,* but it is defined in legal, economic, and political documents and discourse within and outside wildcrafting regions like Appalachia. The person who performs wildcrafting, a wildcrafter, is often a local resident of limited financial means or a hobbyist who spends a certain time each year gathering wild plants for collection or processing.

What makes this craft so important is that when employed properly it is an appropriate technology, but when practiced in excess it can lead to the extermination or endangerment of specific native species. If wildcrafting were confined to targeting or utilizing invasive exotics such as wild chicory (whose roots can be roasted to make a coffee substitute) or kudzu (see chapter 17), it would be a remedial environmental action worthy of immense praise. However, such plants are not the normal targets of the wildcrafter, who deliberately seeks out native plants of greater economic value. Roasted chicory would bring only a few dollars per pound, if a market even exists; wild ginseng root, depending on quality, could bring more than $400 a pound in the ready Chinese market. Wildcrafters may be poor, but they are not ignorant; they know something about economic value.

Wildcrafting is not a new phenomenon. From ancient times indigenous healers were on the lookout for herbs, roots, and plant leaves that, through handed-down cultural and ethnobotanical knowledge, were known for their medicinal value. Some of these plants were processed and ingested as food, some were used to decorate homes or bodies, some were turned into jewelry or baskets. This legitimate art or skill of wildcrafting developed over centuries on every inhabited continent. In modern times scientists have studied wildcrafting in hopes of identifying plants with potential medicinal value.

Wildcrafting can be Earth-friendly, provided it does not endanger species through overharvesting. Rosemary Gladstar, founder of United Plant Savers, says that many wildflowers as well as medicinal herbs and other wild plants are endangered by overcollection. Nurseries and unscrupulous marketers hire immigrants and others willing to work for low pay to gather sacks of moss and wild plants, pot them, and offer them as nursery specimens. Besides American ginseng, which we will focus upon, goldenseal *(Hydrastis canadensis),* lady's slipper orchid (*Cypripedium* spp.), and many other species are being harvested to extinction, and the expanding list includes native cacti as well. Some wildcrafters harvest these plants in haste to beat other greedy foragers. Let's first look at less-threatened wildcrafted edibles and then move on to more economically significant medicinals and ornamentals.

WILDCRAFTING FOR FOOD

Wildcrafting for food, though less common than for medicinals and ornamentals, requires equal skill and technique. The world has at least fifty thousand edible plants, of which most Americans recognize and eat only a few dozen. We include many examples in Appalachia in table 15.1. Foods can be gathered during all four seasons: greens especially in the spring, fruit and berries in the summer, nuts in the autumn, edible roots in the late fall and winter, maple and other saps for processing in late winter, and mushrooms whenever sufficient ground moisture is present. (See chapter 13 for a discussion of the ethics of choosing nontimber forest products.)

Eating wild foods provides variety at little or no cost before spring gardens are highly productive. By eating these wild plants we develop an awareness of the land and seasons. But to eat we must first gather, and so we start by applying the already-mentioned adage "Take a mess, don't make a mess," with "mess" defined as what can be eaten at a single sitting or meal. A modern habit that has crept into the region is to take far more than is needed at a given time just because it is available for the taking. We should remind ourselves of another adage, "Take only what you need and leave the rest."

Greens

Concentrating on gathering escaped or introduced species of greens will allow many of the native ones to continue reproducing. The introduced or escaped varieties are often found along roadsides. Most gatherers of early spring greens never endanger species because plants are small and not yet mature, and expert gatherers only cut the species and quantity that they have learned from their elders to be good and sustainable. Relatively small amounts of greens are gathered so they can be eaten immediately, for greens do not retain their nutritional content when stored. See table 15.1 for scientific name, description, edible parts, location where generally found, and season for gathering.

The dandelion *(Taraxacum officinale)* is a highly tenacious, naturalized plant originally from Europe that spreads to lawns

Table 15.1. Appalachian Wild Edible Plants

Common Name	Scientific Name	Brief Description	Edible Parts	Location	Season
Bracken fern	*Pteridium aquilinum*	large, coarse, erect fronds; claw-shaped fiddleheads covered in silver/gray hair	fiddlehead	dry, open sunny places, woods, and fields	spring
Cane, large (river cane)	*Arundinaria gigantea*	5–30 ft. tall; woody, hollow-stemmed; long, flat grasslike leaves; branched older stems	shoots, seeds	low ground, riverbanks swamps	early spring–summer
Cattails	*Typha* spp.	grow in stands; erect swordlike leaves; stiff unbranched stems; brown cylindrical fruiting heads	shoots, stalks, pollen, roots	marshes, shallow water	all year
Chicory	*Cichorium intybus*	stiff, nearly naked stems; stalkless blue (or white) flowers; dandelion-like basal leaves; milky sap	young leaves, roots	roadsides, disturbed ground	fall–early spring
Chickweeds	*Stellaria* spp.	small plants; slender stems (usually smooth, 1 type hairy); small stalked flowers with deeply notched petals	tender leaves, stems	moist soil, disturbed ground, roadsides	fall–early spring
Chufa	*Cyperus esculentus*	3-sided stem; light green grasslike basal leaves; feathered flower cluster; small, nutlike tubers	tubers	damp sandy soil, disturbed ground	all year
Cleavers (sticktights)	*Galium aparine*	square, weak stems with bristles; narrow, whorled leaves; stalked 4-petaled flowers; tiny, 2-lobed fruit	shoots, roots	thickets, woods, disturbed ground	spring–early summer

Table 15.1 *(continued)*

Common Name	Scientific Name	Brief Description	Edible Parts	Location	Season
Dandelion	*Taraxacum officinale*	sharp, irregular lobes on leaves; hollow milky stems; unbranched conspicuous yellow flower becomes puffy white seedball	young leaves, buds, flowers, roots	lawns, roadsides	early spring–early summer
Daylily	*Hemerocallis fulva*	orange unspotted blossoms (open 1 day); leafless flower stalk; swordlike basal leaves	shoots, buds, flowers, tubers	disturbed ground	all year
Dock	*Rumex* spp.	coarse leaves often with wavy margins; dense heads of small greenish flowers or brown seeds with wings	young leaves	fields, disturbed ground	spring
Evening primrose	*Oenothera biennis*	(biennial) 1st year–low rosette of leaves; 2nd year–red-stemmed stalk, alternating leaves; yellow, short-lived blooms	young leaves, roots	dry soil, roadsides, disturbed ground	fall–early spring
Greenbriar(s)	*Smilax* spp.	green-stemmed, prickly climbing vines; parallel-veined leaves; greenish flowers and small berries	shoots, young leaves, roots	woods, thickets	all year
Jerusalem artichoke	*Helianthus tuberosis*	tall sunflower; broad rough leaves; hairy stems; upper leaves alternate; large tubers	tubers	disturbed ground, damp thickets, fields	fall–early spring

Table 15.1 (continued)

Common Name	Scientific Name	Brief Description	Edible Parts	Location	Season
Kudzu	*Pueraria lobata*	30–100 ft. trailing or climbing vine; downy young stems; large leaves with broad oval leaflets; purple/violet flowers	roots, young leaves	thickets, bordering woods	all year
Lamb's quarters	*Chenopodium* spp.	erect, many-branched variable leaves (upper–narrow, toothless; lower–broadly toothed); white stems and underside	tender leaves and tips, seeds	disturbed ground	summer–early winter
Milkweed	*Asclepias syriaca*	stout, downy plant; milky juice; domed clusters of flowers (purplish, buff, or white); warty, pointed seedpods	shoots, leaves, buds, flowers, young pods	dry soil, fields, roadsides	spring–summer
Mint(s)	*Mentha* spp.	aromatic; square stems; paired leaves; clusters of small lipped flowers	leaves	wet places, damp meadows	summer
Mustard(s)	*Brassica* spp.	broad, deeply lobed lower leaves; slender seedpods; terminal clusters of 4-petaled yellow flowers	young leaves and seedpods, buds, seeds	disturbed ground, fields	early spring–summer
Nettles	*Urtica* spp.	distinctive stinging hairs; erect unbranched, paired, toothed leaves; tiny green flowers in slender forking clusters	shoots, young leaves	roadsides, disturbed ground, thickets	spring–summer

Table 15.1 (continued)

Common Name	Scientific Name	Brief Description	Edible Parts	Location	Season
Plantain	*Plantago major*	broad, ovate, heavily ribbed basal leaves with troughlike stems; slender-headed greenish white flowers; leafless stems	young leaves	lawns, disturbed ground	early spring
Pokeweed	*Phytolacca americana*	large leaves; smooth reddish stems; long-stalked flower; clusters of glossy purple-black berries	young shoots *only*	roadsides, fields, disturbed ground	spring
Solomon's seal	*Polygonatum biflorum*	parallel veined, alternating leaves; arching stem; paired greenish yellow bells; blue-black berries dangle beneath leaves	young shoots, roots	woods, thickets	all year
Sorrel(s)	*Oxalis* spp.	grayish-green cloverlike leaves; long leaf stalks on thin stems; yellow or violet flowers	leaves, stems	fields, roadsides	spring–late summer
Spring beauty	*Claytonia virginica*	small white to pale rose 5-petaled wildflower; 2 thin, pointed leaves on each thin stem; irregularly shaped tubers (corms)	corms (tubers)	moist woods, rich soil	early spring–spring
Sumac, staghorn	*Rhus typhina*	shrub; smooth green bark, raised cross streaks; alternating compound leaves; dark red berries covered in bright red fuzz	fruit	upland, old fields and openings	summer

Table 15.1 (continued)

Common Name	Scientific Name	Brief Description	Edible Parts	Location	Season
Violet(s)	*Viola* spp.	slender, wiry branching stems; oval or heart-shaped leaves; 5-petaled white, yellow, or blue flowers	young leaves, flowers	damp woods, meadows	early spring–spring
Waterleaf (Shawnee)	*Hydrophyllum* spp.	rosette of basal leaves, marked with lighter green; radiating clusters of 5-petaled white or violet flowers	young leaves	woods, most soil	early spring
Wild asparagus	*Asparagus officinalis*	feathery bunches of threadlike green branchlets in brownish, scalelike leaves; tiny greenish-yellow flowers	young shoots	roadsides, fields, fencerows	early spring
Wild carrot	*Daucus carota*	(biennial) hairy-stemmed; lacy flat-topped flower clusters, often a single purple flower in center; stiff 3-forked bracts	roots	fields, disturbed ground	fall–early spring
Wild ginger	*Asarum canadense*	large, paired heart-shaped leaves; 2 stout, woody leaf stalks; single 3-lobed, bell-shaped reddish-brown flower	roots	rich woods, near rocks	early spring–fall
Wild lettuce	*Lactuca* spp.	tall, leafy, dandelion-like leaves; long, loosely branched flower clusters; milky bitter sap	young leaves, developing flowerheads	thickets, clearings, roadsides	spring–summer

Table 15.1 (continued)

Common Name	Scientific Name	Brief Description	Edible Parts	Location	Season
Wild onion/garlic/leek	*Allium* spp.	grasslike basal leaves; small 6-petaled flowers; smells like onion or garlic	leaves, underground bulb, bulblets	fields, roadsides, meadows, open woods	all year
Morel	*Morchella esculenta*	(mushroom) cap is pale tan to gray, spongelike; blunt, cone-shaped; lower end fused to stem; deeply pitted with whitish ridges	cap, stem	moist woods, orchards	spring
Puffball	*Calvatia gigantea*	large, smooth, white, globular mushroom grows from ground; interior flesh pure white with no rudimentary stem or gills	fruiting body	disturbed ground, open places, pastures	late summer–fall
Fruits: Crab apple, mayapple, pawpaw, wild cherry, wild grapes					
Nuts: Acorn, American chestnut, American hazelnut, hickory nut, walnut					
Berries: Black raspberry, blackberry, dewberry, elderberry, juneberry, red raspberry, wild strawberry, wild blueberry					

SOURCE: ASPI, *Appalachian Edible Wild Plants,* Technical Paper 23 (Mount Vernon, KY: ASPI, 1993).

NOTE: This chart is not inclusive, nor is it meant to be a substitute for a field guide. It is provided as an illustration of the great variety and availability of wild edibles in Appalachia and may help you reexamine plants with which are you are already familiar. Several plants on this list are now threatened or endangered and should be tasted but not harvested.

and roadsides. It is the basic green, so take all you want, for it is early (we see blooms every month of the year) and grows first among the fallen leaves. Cut this and most other greens just above the roots so that the bunch comes out as one green clump, which can easily be shaken free of soil. Dandelions can be eaten either green (when tender), wilted with hot oil along with onions and potatoes, or cooked as a potherb. The dry root is a nutritious coffee substitute.

Another Appalachian green is pokeweed, a tall, bright red plant. It is regarded by some as a poison, but Appalachians know the early leaves make "sallet" and the red coating of young stems can be peeled off and the poke fried or prepared like asparagus. I eat poke greens throughout the growing year, discarding the first boiling water from mature plants. What about the berries? Lucille Steward, an expert Kentucky greens gatherer, says a Cherokee taught her to eat one berry a day for arthritis (swallowed whole, for the poison is in the seed) from shorter and younger stalks. She takes her berries to eat from a year's supply stored in the freezer. I follow her example and gather berries in the autumn.

Berries

Picking as many blackberries as you wish does not harm the total supply, unless you attempt to destroy the patch in the process. Good wildcrafting means knocking down older briars but saving the green young ones, which will bear next year. Selectively knocking down growth can actually assist the plants by making the coming shoots bear well the following year. Blueberries and elderberries grow at reasonable heights and lack thorns. Pick to your heart's content.

Nuts

Beat the squirrels and varmints and don't harm the trees in the process. This may apply more to hazelnuts, which grow on rather fragile bushes, than to walnut and hickory trees. For chestnuts, remember that the spiny hulls of the sound nuts open easily, but not the unsound ones. Wear gloves when hulling walnuts or leave them on the roadway for cars to run over. Remember that the squirrels are on the lookout in autumn.

Roots and Bark

Caution: this is more problematic wildcrafting, for trees, shrubs, or other plants can be damaged in taking roots and bark. Generally, do not dig roots unless you compensate in some way, such as reseeding or replanting tree seedlings to allow the rest of the species to thrive. Take a little sassafras, but not too much. Sassafras can be replanted.

Mushrooms

More mushroom species—three thousand to five thousand—grow in the Southeast than anywhere else in America. How many of these are edible is unknown. These nutritious fruiting bodies of fungi are generally safe from extinction because many people fear the toadstools and shy away from the rest. Experts are few, and markets for unusual mushrooms are weak compared to shiitake or the other, relatively few, commercial types. Residents do recognize morels like *Morchella esculenta, M. crassipes* (thick-footed morel), *M. elata* (black morel), and *M. conica*. Other popular types found in Appalachian woods include oyster mushrooms, chanterelles, shaggy main, velvet foot, and hen-of-the-woods. Individual gatherers do little damage either to the substrate medium (rotten logs) or to the spawning spores. However, mushrooms live in a symbiotic relationship with a healthy forest environment that is currently being threatened. Damaging the upper four inches of forest floor through acid rain or unsustainable forest harvesting will threaten future mushroom wildcrafting.

GINSENG AND OTHER MEDICINALS

Medicinals such as ginseng are at the heart of economic wildcrafting. Some people outside the regular economic system get their Christmas money from digging wild plants (leaves and/or roots) and selling them to a variety of wholesalers and merchants, who pay in cash. The custom of earning money from ginseng is time-honored, and wildcrafting is a regional way elders bond with spry youngsters with visions of sugar plums (what Appalachians call persimmons, which can be picked off the tree at Thanksgiving). While the focus is on threatened wild ginseng, a similar but generally less urgent argument can be made for dozens of other medicinals, except that market prices are far lower.

Wildcrafters turn quickly to the beautiful, tall, green, Appalachian foremost medicinal, American ginseng *(Panax quinquefolium)*, a cousin of the much-treasured Asian ginseng *(Panax ginseng)* (see figure 15.1). Asian ginseng has occupied a central place in Asian medicine for several millennia and brings hundreds of dollars a pound on that market today. The loss of the Chinese temperate hardwood forest centuries ago doomed the prized wild Asian ginseng, for it only grows naturally in forested areas.

The Cherokee have enjoyed a long medicinal relationship with the plant they refer to as Grandfather *yvwi usdi,* or "Little Person." When the Cherokee dug ginseng, they did not dig the first three plants they found. When the fourth plant was dug, they would place a pebble, bead, or even a marble in the earth in gratitude to the plant spirit.

With the demise of naturally grown Chinese ginseng, eyes turned to North America. Appalachia has been known for ginseng root from Colonial times; Daniel Boone shipped bags of the root across the mountains in the eighteenth century, and John Jacob Astor's clipper ships made ginseng one of America's first major exports.

Today, lower-grade, artificially covered, cultivated ginseng, principally from Wisconsin (bringing about \$8 to \$10 a pound),

Fig. 15.1. Ginseng patch at an undisclosed Appalachian location

accounts for 90 percent of America's 1.5-million-pound export business. But the wild root is still so prized that illegal wild-crafters are endangering the species to the extent that it may not survive this decade. Smaller and smaller roots are being dug, and federal agencies that regulate exports of endangered products. have reacted with alarm. Export of ginseng root less than ten years old is prohibited, but the rule is virtually unenforceable.

Another current danger to wild ginseng is the proliferation of wild turkeys. These flocks scratch their way through Appalachia and consume all the bright red fleshy ginseng seeds in midsummer. Turkeys differ from pheasants, which pass the seed intact through their digestive tracts; the turkeys have gizzards that grind, crush. and ingest the ginseng seed material.

The growth of ginseng can be enhanced in cultivated form using artificial cover or through semicultivation under forest cover. ASPI has endorsed a method adapted by ginseng grower

Syl Yunker, "responsible wildcrafting." Promising, well-drained hardwood forested sites are selected, sparingly sown with ginseng seed, and protected from poachers. Within a dozen years excellent-grade organic wild ginseng (as opposed to cultivated crops with heavy pesticide treatments) is dug either green or dried. Today, green root can be air-freighted to Asia without moisture loss. A virtually wild crop is indistinguishable from wild ginseng and brings a far greater return than most medicinals. Because the Chinese regard ginseng so highly for treating about two hundred ailments, this plant's potential to save Appalachia's forests is enormous. The market is well established, so forest owners are strongly encouraged to grow the plants and also consider selling the highly potent leaves, a practice that allows roots to grow and mature with time.

Wild ginseng contains an assortment of chemical adaptogens that may work synergistically, like a chemical buffet, to calm the body and allow stability, giving the body a chance to heal. There is no question that the medicinal has proven to be effective; the question is just how effective and for what ailments. Laboratories are striving to isolate the chemical ginsenosides in the plant root and (thanks in part to ASPI's extractive efforts) the leaf gathered from wild ginseng roots in Kentucky and North Carolina. The increased focus on leaves will allow growers to get some cash return while they await mature ginseng roots. Researchers hope to synthesize the active ingredients, but wild ginseng growers are not deeply concerned, because the natural product is so difficult to duplicate and the ginseng market is expanding rapidly as Americans discover ginseng's health benefits.

Hopefully, the U.S. Fish and Wildlife Service will use the CITES Treaty, signed by the United States decades ago, to protect endangered species from international commerce. But more will have to be done to protect wild ginseng on private landholdings from trespassers who would steal every plant in sight. Efforts are being made by a number of us to encourage the government to require a marketing card that would prove that the crop grew on the seller's property. More law enforcement efforts are now being undertaken to curtail widespread poaching, especially since some of the poaching takes place on federal land. Small-time wildcrafters need not worry if they have returned seed from mature harvested plants to the soil and have permission to dig on specific public or private lands.

A danger with virtually wild or other forest-cultivated types of ginseng is that these plants might cross-pollinate with wild ginseng. Ginseng does exhibit genetic diversity, with regional varieties adapting to local climate and soil conditions. Seed stock from outside the region may not reproduce as successfully as native stock. While ginseng is primarily a self-pollinating plant, some cross-pollination is provided by sweat bees and hover flies and could result in some genetic depression. A simple solution is to start virtually wild plots from distant seed in areas where there is now no wild ginseng.

Ginseng is only one of more than a hundred commercial medicinals found in Appalachia. A review of all of them is not possible here. Each has its champions, marketing agents, and wildcrafters. With such ginseng companion plants as goldenseal, blue and black cohosh, bloodroot, and wild ginger, the ignorance of potential trespassers is the best protection against overharvesting. Widespread publicity will only threaten these plants. The market for many of the herbal medicinals is quite volatile, and in many cases the cultivated varieties are believed to be about equal in value to the wild plants, a situation that contrasts with that of ginseng.

ORNAMENTAL AND OTHER USES OF WILD PLANTS

The options here are so many that one hesitates to launch into this subject. A general rule is to harvest exotics and especially invasive species in whatever quantity is desired. Kudzu fiber makes excellent ornamental objects. The use of corn husk products can be encouraged, since the fiber is easily found, but the use of wild native fibers or vegetative matter should be limited.

Flower gathering is a dangerous wildcrafting area, since this could truly endanger native orchids, trillium, and other species. Hikers may spy a beautiful flower and decide to dig it up without regard to whether such a plant could live where it is taken. A simple rule is: do not pluck or dig up a native flowering plant. This is against the law in a number of countries and on public lands in the United States. Photograph and describe a plant and do research to find out what it is, but do not cut or trim or dig out the roots. Amazingly, many otherwise Earth-friendly folks disobey that rule. Some would argue that wildflower seed can be collected after the blooming period, but it is best to leave the collecting to botanical experts, for many seeds will grow only in highly specific soils and weather conditions. For bouquets, consider either cultivated species or the plentiful exotics such as wild sunflowers, chicory, ironweed, daffodils, and daylilies.

As for taking stems, vines, and branches for ornamental use, stick to the exotics. The kudzu vine can be taken by the carload, and it is right on the ground in many places. The proliferation of Oriental bittersweet (*Celastrus orbiculatus*) makes it a top choice for wreath making and basket weaving. Golden bamboo *(Phyllostacyhys aurea)* and other invasive bamboos (*Phyllostacyhys* and *Bambusa* spp.) can be used for fishing poles, plant trellises, woven privacy screens, and many other creative craft projects. Again, wildcrafters should gather the exotic species or the plant pests rather than native materials that are often under stress. Kudzu and other exotic vines have been used for centuries in Asia for ornamental and practical baskets, seats, wreaths, dolls, and mats. Using coarse wild materials for fibrous articles has been an Appalachian specialty, and articles made from wildcrafted materials find their way into craft shops and centers throughout the region. Efforts should be made to encourage crafting with invasive species. The enormous creativity of crafters is limited only by their imaginations and could be the subject of a book in itself.

CHAPTER 16

Constructed or Artificial Wetlands

Water is one of our precious resources and we are putting our liquid and solid wastes into our streams and thus destroying this resource. Is there a way to treat our sink and shower water in places where the soil is tight or mostly clay or where there is bedrock close to the surface? Also what can be done for the many homes that have little space for the septic tank and leach field? The sink and shower water can be treated nicely in a constructed wetland coupled with the use of a composting toilet. In the constructed wetland the liquid moves slowly through it as it would in a natural wetland or marsh at the shore of a river or a lake. As the liquid slows down and the particles settle out, the bacteria and fungi break down the food and soaps that have entered into the liquid. They convert the food and soaps making them available to the plants, which finish cleaning the water. These plants beautify the yard and are kept watered even in dry seasons.

Jack Kieffer, research scientist, ASPI

Eastern America has suffered from a sizable loss of its natural wetlands and bogs in the past two centuries. Some estimate that as much as 60 percent of American wetlands were destroyed during the period of major European settlement. The destruction has occurred through ditching for agricultural drainage, road building, and other development projects. During much of this time settlers thought of wetlands as waste areas, prime candidates for human conquest. Thus the swamps were drained and cleared to rid the landscape of snakes and mosquitoes and other undesirable varmints and to render it fit for human settlement. During recent hurricane seasons the loss of Gulf Coast wetlands has had severe economic and ecological consequences, as most Americans have learned.

This transition has not been as severe in mountainous regions like Appalachia as in lowland areas because the mountains have far fewer wetlands. But while they are less numerous, natural Appalachian bogs are ecologically unique and valuable. In recent years, environmentally conscious people have come to appreciate the natural capital of wetlands as rich ecosystems that attract a multitude of wildlife. The term "wetland" now has come to mean a place from which a rich variety of wildlife springs, where wastewater is purified, and where multitudes of birds, mammals, reptiles, and other wildlife seek protection and breeding opportunities. Within the land conservation movement there is a strong thrust to restore wetlands in many coastal parts of the United States; the U.S. Forest Service, the National Park Service, and various state governments are undertaking measures to restore wetlands in upland forested areas as well.

Reversing wetland losses in our own backyards and communities would have a number of beneficial effects:

These constructed wetlands can be built in difficult terrain and on moderate to steep slopes and rocky uplands where it is usually impossible to build traditional leach fields.

They can be constructed at low cost by everyday do-it-yourselfers following simple rules and using mostly locally available materials.

Wetlands can efficiently process in a natural way domestic wastewater such as gray water from washing dishes and bathing.

Wetlands can be coupled with homes equipped with dry composting toilets (see chapters 25 and 26).

Wetlands have the potential to be transformed into beautiful flower beds with an abundance of native species.

The principle of wastewater purification through a natural biological treatment method is quite straightforward. Wastewater from a typical residence enters the wetland cell at one end and comes in contact with aquatic-type (succulent) plants, which are planted in a 1-foot-deep gravel medium. This medium is in a 4-foot-wide trench that has been lined with a plastic liner and to which is added 2 inches of mulch. Plants are set in the mulch with their roots in the gravel. As oxygen from the air penetrates the surface mulch, beneficial microorganisms and fungi thrive and reproduce in the system. These organisms attach to plant roots and rock media, where they, together with the plants, utilize organic matter and nutrients from the wastewater as food and fertilizer. All the while, large amounts of water from the wetland may be lost through evapotranspiration via the aquatic plant stems and foliage. Thus wetlands become natural water treatment facilities.

The resulting effluent is purified to a very high degree. In a way, much depends on the volume of water coming from the waste source. A rule of thumb is that there should be 1 cubic foot of artificial wetland for every gallon per day of wastewater effluent, or a 120-square-foot cell, 1 foot deep, for a single-bedroom house. However, homes with a dry compost toilet so dramatically reduce water consumption that wetlands built according to regulations set for flush toilets can be overbuilt. Where water conservation is practiced, the wetland may be far smaller, but some states (Kentucky, for instance) specify certain sizes per number of bedrooms or inhabitants. Even when one uses less water, it still pays to overbuild the wetland. Our ASPI constructed wetlands have never had outflow from the basic cells even when rainfall was very heavy. Furthermore, the more conservation conscious may use some gray water for greenhouse watering or other uses, reducing still further the amount of water going into the wetlands. On the other hand, during times of drought some wetland operators may have to add water to sustain wetland plants.

The ASPI Appalachian constructed wetland (see figure 16.1) is built on a steep slope in a terraced fashion below the solar house near Livingston, Kentucky. The shape is certainly different from flatland wetlands, but the principles used in construction are the same as for the seven thousand naturally occurring wetlands in the level Ohio–Mississippi river valley in western Kentucky. In siting a constructed wetland, many prefer a prominent location because wetlands are beautiful and worthy of notice. Often one sees the wetland near the front door of the house. The final choice of site depends on the proximity to the rest of the facility, the quantity of water to be processed, natural topography, access to hauled construction materials (mainly gravel), and placement with respect to the building served. A proper length-to-width ratio of wetlands should be maintained; this depends upon the water volume entering the field (e.g., 4 feet by 90 feet for a standard three-bedroom, two-bath house).

Regulatory agencies have overflow field requirements that must be considered in wetland construction. Potential wetland installers must check on local and state requirements. Agencies often require that a septic tank be inserted after a flush toilet and prior to the wetland proper, and most have an overflow requirement of a standard or modified leach field or a French drain (a ditch with gravel and perforated drainpipe to allow slow percolation of any possible overflow). In Kentucky, for instance, for a two-bedroom house, the size of the cell should be 240 square feet (4 feet wide by 60 feet in length), and the length of the laterals should be 200 feet. Final sizing of the constructed wetland will depend on the soil conditions, the type and volume of water flow, and the particular design features of the total system.

Fig. 16.1. ASPI contour wetland

Inhabitants of the lowlands and plains can lay out rectangles for wetlands, but that's not easy in mountainous terrain. Constructed wetland shapes may have the grace and beauty of the terrain itself. In Appalachia, level land is at a premium, and roads and trees tend to get in the way of the small amount of green space that some people have around their homes. The answer for the mountain constructed wetland is to curve the cells on contour with the hillsides, provided the curves are not so sharp that the long strips of plastic liners can't be properly fashioned as needed under the gravel bed. A border for the terraced wetland can be constructed of railroad ties or native materials such as black locust logs to hold the edge of the liner in place. Stone or rot-resistant wood can be used for a facing, depending on available materials, the artistic skill of the builder, and the budget of the wetland owner. Some Appalachian wetlands conforming to the curvature of the terrain have the potential to be works of art folded into the earth.

HIGHLIGHTS OF CONSTRUCTION

The building of wetlands in rural America is a reclamation project at the grassroots level, even though it may be at a distance from where a wetland has been destroyed. The actual project is usually on a small scale and thus does not require major pieces of equipment or great outlays of money, even though wetlands are built in some cases for entire towns. One urban example is in Mount Angel, Oregon, about an hour south of Portland on Interstate 5. There the municipally constructed wetland serves three thousand people and has been working very well for years. Other constructed wetlands can be found at interstate rest stops in Tennessee and near commercial buildings at many rural locations.

For a small-scale wetland, the construction can be below grade with hand tools or a backhoe, or equal grade with a berm of construction materials. The more creative the curved shape, the greater the need for hand tools and the less the need for heavier equipment. In terrace building, the use of hand tools has been the choice for some of us because bringing in mechanized digging equipment would seriously disturb the roots of nearby trees and damage adjacent wooded areas. Healthy trees are ideal for terraced wetlands. A slight slope (1 inch to 100 feet) to allow the water to run by gravity can be determined by a standard transit or by a level. A second unlined cell should be about 12 inches lower than the first and at the same slope for gravity flow as the initial cell.

The wetlands can have two beds, one lined with PVC, polyethylene, butyl rubber, or comparable materials of at least 24 millimeters in thickness, with 30 millimeters preferred. The liner must be protected from sunlight before and after installation, and care should be taken to cover the edges and to lay the liner without puncturing it. An impermeable clay liner could be used as a natural liner in place of the artificial liners, but that is not always available. Care needs to be taken to tamp the disturbed subsurface area properly to ensure that it does not leak.

In building constructed wetlands a variety of materials can be used as fillers: clean road gravel (about ¾ inch in diameter), pea gravel, or river or creek gravel. If rocks are in the exposed soil surface, they can be removed and sand added as a layer before installation of the liner and gravel to cushion the area. Over the gravel should be placed a 2-inch layer of mulch to protect the plants from summer heat, prevent odors, and give an aesthetic finish to the cell. This mulch should be made inactive through kiln or air-drying so as not to damage the aquatic plants. It should be coarse enough not to clog the gravel through which the water is to flow. This mulch should be spread evenly as an exposed top layer. (See figure 16.2.)

Besides the liner and filler, the major expense for the project is the PVC pipe (about 2 to 3 inches in diameter, depending on water volume) that connects the house to the cell and between cells. Pipes should be sealed tightly to liners at the point of penetration of both inlet and outlet pipes, and a tight-fitting gasket should be installed; the pipe and liner should be held snugly with two plumber's clamps. The inlet and outlet headers should be made from the same diameter of pipe. One-half- to three-quar-

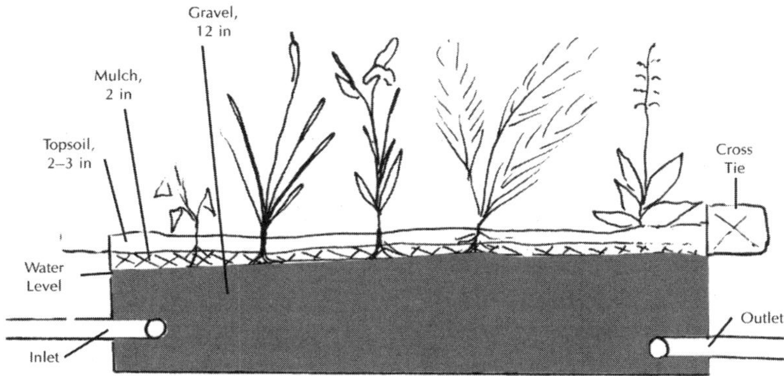

Fig. 16.2. Cross section of an artificial wetland

ter-inch holes should be drilled about a foot apart in the header. The inlet header should be placed across the width of, and perpendicular to, the elongated wetland so that the wastewater will move evenly through the length of the cell. The outlet header could be the same size pipe or slightly smaller, with comparable holes drilled in it. The holes should be placed in the direction of the water flow. Lateral distribution piping to receive any possible outflow of the wetland is required by regulatory agencies.

A plugged clean-out located above ground should be installed at the inlet area; this should be capped and available for emergency use. On the outlet end should be placed a level controller consisting of a PVC elbow pipe. This can be manually swiveled to allow water to stay or be emitted from the primary cell to the French drain or leach field. This should be protected by a box built at the edge of the wetland bordering material. (See figure 16.3.)

WETLAND PLANTS

Wetland plants should be carefully chosen, for these are the wetland's workhorses. When and where possible, go native. Some may wish to divide the wetland into lined and unlined portions, with the first planted with more aquatic plants. Most ecologically minded operators regard an established polyculture as more desirable than a monoculture. Much depends on the native plants of the region. But wetland conditions are so controlled and bordered that even noninvasive nonnatives could be considered for the more aquatic species. We have observed a great variety of flowering and other plant choices for both heavy and light evapotranspirator agents.

Some regard their small domestic property holdings as so valuable that much of the plant life should be edible. Common cattails have edible shoots. Certain of the many canes or wild reeds, including the one species of native American bamboo, river cane, which is found in Appalachian river bottoms, can be

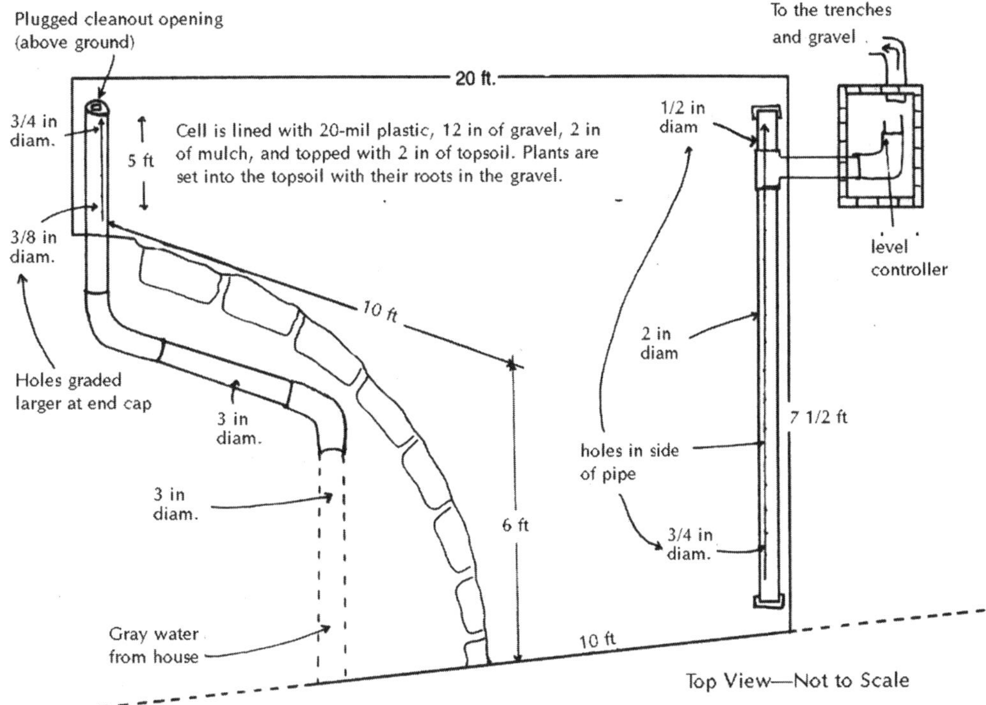

Fig. 16.3. Schematic of ASPI's artificial wetland

used for decorative and building materials and, in some cases, for food. The "edible" label also applies to plants that nourish birds and other wildlife. It is possible to add wetland plants that attract butterflies, such as bee balm (Oswego tea). Other possible plants include the marsh milkweed, pickerelweed, blue water iris, and sweet flag. Plants mentioned in chapter 28 are also good candidates for Appalachian constructed wetlands.

Other aquatic plant choices might include cardinal flower, water primrose, columbine, Virginia meadow beauty, spearmint, tiger lily, crooked stem aster, lizard's tail, black-eyed Susan, sensitive fern, cinnamon fern, royal fern (for shade), calla lily, elephant ear, ginger lily, turtlehead, monkey flower, water willow, marsh blue violet, mist flower, and nodding bur marigold. Many of these are good providers of nectar for honeybees and thus extend the value of the edible landscape to important pollinators.

ASPI used impatiens purchased from the local greenhouse, and this flower proved to be quite satisfactory as a wetland candidate. However, the plant is an annual for the most part, is an exotic from the semitropics, and, if commercially grown, is expensive considering the number needed to cover the wetlands. Another plant from the *Impatiens* genus and the balsam family is jewelweed, a lovely yellow- or orange-flowered plant also known as wild touch-me-not, which grows as a volunteer native plant throughout Appalachia. This plant has a high succulent or water-absorbing potential because of its size, and an added advantage is that the plant juice serves as a soothing medicinal ointment for skin irritated by allergic reactions to poison ivy, insect bites, and other conditions. When we finally appreciated its power as a wetland provider, it became our favorite. Jewelweed can be allowed to grow throughout the wetland space, depending on how someone wishes to sculpt the flower garden in the overall wetland design.

A special instance of a gray-water flower bed is worth mentioning: the NutriCycle Systems Graywater Flower Bed (see Resources). The gray water is automatically and safely delivered to the root zone of vegetation in flower beds designed aesthetically into the landscape. When absorbed by the vegetation, the organic matter and nutrients found in gray water are recycled to the land-based food chain, and all water not transpired is recycled to the groundwater without health hazards or pollution. For this application a compost toilet is required. The compost toilet reduces water usage significantly, and its compost tea, if formed, is not added to the gray water. A septic system is not allowed as part of the system because pathogens multiply in the liquid retained in the septic tank and the oxygen in the gray water is depleted.

The above requirements regarding the compost toilet and septic system are needed so that the gray water that is dosed in the garden will still contain oxygen that will help promote plant growth. The gray water is collected in a dosing tank equipped with either dosing siphons or pumps; this arrangement allows half of the emitted gray water to water half of the garden every other day. From the dosing tank the gray water flows into troughs that have side walls but are open to the soil and roots of the plants. The plants are grown on either side of the trough and use the water and nutrients supplied by the gray water. The soil bacteria and fungi break down the food particles and soaps so that they are ready for the plants to use.

CHAPTER 17

Land Reclamation with Native Species

Our lands in eastern Kentucky and neighboring Virginia have been hard hit by both the coal strip mining outfits and the loggers. The tree-covered mountains we knew since being youngsters are quickly skimmed off or cut over and turned into ugly brown patches of land that can easily wash away. And it did just that during some of the terrible floods we have seen since 1977. Bottomlands get covered, homes and roads wash away, creeks silt over, and the fish are gone. These sore sites could be made better by government enforcement of existing laws. Will it happen in my lifetime? I really don't know.

Becky Simpson, Cranks Creek Survival Center,
Harlan County, Kentucky

Appalachia has suffered much from various forms of land disturbance, which have included farming on steep slopes, overharvesting of timber, and the surface mining of coal. As a result, the land is prone to severe flooding because of the removal of the sponge of the forest cushion, landslides, drops in the water table, the fracturing of aquifers, major erosion, and siltation of streams and rivers. The life of mountain communities has also been disrupted in these land-disturbed areas because Appalachian people and the land on which they reside are so intimately connected. To hurt the Earth is to hurt its people.

Land disturbances have become increasingly noticeable in recent decades. Besides highway construction, grading of building sites, unsustainable timber and fuel-wood harvesting, and surface mining of coal are such practices as development of urban industrial zones, construction of parking areas and airports, and development of artificial lakes. The pace of disturbance has not slackened in the twenty-first century, for the chip mills and the demand for timber call for the felling of more trees, and surface miners have turned to slicing off the tops of mountains and filling in the valleys using gigantic earthmoving equipment. Though mountaintop removal is an environmental travesty for given locations, the rate of new surface-area disturbance has recently decreased due to a reduction in contour mining. The result is a greater intensity of disturbance on less land, but a practice that leaves an indelible mark on Appalachia.

Overall, about one-tenth of the land surface in central Appalachia has undergone major disturbance. In the face of such massive disruption, what can people do? Some regard the disturbances as necessary evils for the sake of jobs. Others champion deep-mining alternatives and say the major land disturbances should never have occurred. But in spite of the efforts of activists since the 1960s, land is still being damaged. The economic and political forces hold sway and bring together a strange alliance of corporations, road lobbyists, political parties, and laborers to fight for the necessity of continued land "development," all at the expense of the forest cover that is absolutely needed on such fragile land. Can the disturbance be stopped, slowed down, or mitigated, or is our power only to be directed at reclamation efforts (such as the one shown in figure 17.1) and healing wounded land? Stopping land disturbance is like trying to stop a mighty flood. However, healing through Earth-friendly means is empowering in a mysterious way, though we should not let our positive steps serve as an excuse for continued unchecked disturbance of Appalachian land.

We must think like medical personnel bent on using our time efficiently. Wounded Appalachian lands can be divided into three categories: partly reclaimed, or lands partly cared for using exotic plants or improper reclamation practices; currently disturbed,

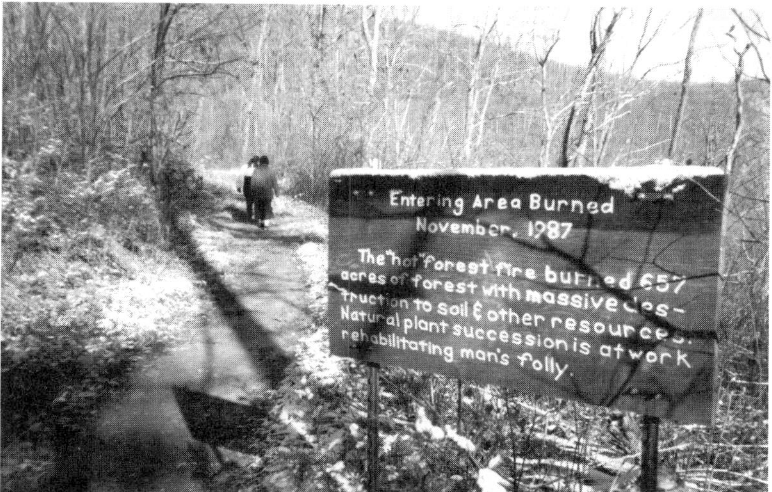

Fig. 17.1. Initiating reclamation work in burnt-over forest in eastern Kentucky in the late 1970s

or lands where some incomplete reclamation has been done because of federal and state regulations; and needing attention, which includes abandoned or forsaken lands that have not been covered under any regulations and continue to erode and degrade. By reclamation, we mean curbing land degradation and starting the natural healing process using native species as vegetative medicine.

EXOTIC SPECIES

The denuding of land in Appalachia started with the advent of European settlement in this region. However, the destructive nature of such a practice has only been recognized in the last century, especially after the introduction of heavy earthmoving and resource-extractive equipment. Generally, a cut-over forest would include many saplings and understory growth that could spring back rather rapidly following the departure of ax-wielding loggers and oxen. Disturbances became more pronounced in the nineteenth and twentieth centuries as land developers used heavier machinery and reached wider areas. It became evident that this fragile steep-sloped land, subject to heavy rainfall, would soon be plagued by floods and major soil erosion if some form of vegetation were not quickly restored. Often the early developers looked beyond native plants for vegetative cover that could proliferate rapidly and in the process return some nutrients to the soil. Kudzu and autumn olives were among the species that were planted. This practice had a major flaw: a lack of control over the rampant spread of these exotic species. Just as the original damage was done with disregard to ecological principles, the early reclamation efforts, both for road cuts and for surface mining, unfortunately were not always based on selecting native species adapted to the Appalachian bioregion.

For example, the multiflora rose *(Rosa multiflora),* a native of Asia that was brought to the United States originally in the 1800s for use as rootstock for grafted ornamental roses, was promoted from the 1930s until the 1950s by the U.S. Department of Agriculture as a "living fence" for windbreaks and erosion control. Birds enjoy the rose hips and thus spread the plant to a wide array of climatic and soil conditions. This seemingly harmless rose is quite capable of choking out native plant life. Persistent mowing can rid lands of the pest. Likewise the invasive tree of heaven *(Ailanthus altissima)* is found now throughout central Appalachia in sunny, disturbed areas and wastelands. Cutting and adding salt to the base of this tree is needed to eradicate it. Japanese honeysuckle *(Lonicera japonica)* was propagated in America for landscaping and erosion control. It is widely known for its fragrance and is as bothersome as its bushy cousin coral honeysuckle (*Lonicera sempervirens*).

Kudzu *(Pueraria lobata)* first arrived in the South from Japan at the New Orleans Exposition in 1884; it was seen as an exotic ornamental with sweet-smelling flowers and lush foliage. At the turn of the twentieth century a southern agricultural association promoted it as fodder for cattle. In 1927 a Georgia promoter declared, "Cotton isn't king here anymore; kudzu is king." By 1934 the U.S. Soil Conservation Service introduced kudzu as a species for use in eroded ravines and for reclaiming road and railroad cuts. About that time academic research at the University of Georgia was directed to find hormones to make kudzu grow faster. After the Second World War the truth began to dawn: kudzu was an uncontrolled menace, growing rapidly beyond intended areas, climbing over buildings and large trees. This exotic species was an invasive. By 1993 the cost of fighting kudzu was estimated at $50 million annually.[1]

Granted, the plant has benefits or it would never have been introduced. Kudzu has a potential to regenerate the fertility of nutrient-depleted soils. As a legume it has a symbiotic relationship with *Rhizobium* and can achieve nitrogen fixation in soils. Besides, it is a green manure, will outcompete weeds and return nutrients and biomass to soil, and makes a fine mulch and an animal fodder that is nutritionally comparable to clover and alfalfa, rich in vitamins A and D, calcium, and protein. The uncontrollable nature of kudzu was only gradually recognized, for the Japanese farmers were far more adept at control measures than were Americans. Thus the first of America's experiments with revegetation of damaged sites was a miserable failure. As a postscript, kudzu now chokes out native species in virtually all states in the Southeast. It covers over 7 million acres of land, more than the surface area of the state of Maryland.

Autumn olive *(Elaeagnus umbellata)* and Russian olive *(Elaeagnus angustifolia)* have been planted throughout Appalachia, especially on strip mine reclamation sites. These shrubs, which can reach 20 feet in height, have been set out to control erosion, return nitrogen to the soil, and provide food for wildlife. Birds enjoy the abundant small red drupes that mature in early autumn; they scatter seeds widely, contributing to making these species worrisome invasives that compete for open, well-drained

---

1. Kurt E. Kinbacher, "The Tangled Story of Kudzu," *Vulcan Historical Review* 4, Spring 2000.

land. These two tree species fortunately are not shade tolerant and will not spread in already heavy forest cover. However, the rugged autumn and Russian olives must be cut and cleared by repeated mowing and grubbing out all of the root parts from the area. There's work to be done!

Privets are popular imported ornamentals that come in many varieties and are regarded as "privet hedges" by landscapers. But they are considered invasives by many environmentalists. Furthermore, they are poisonous and are not evergreen in the North, so they are attractive for only a portion of the year.

PAST AND CURRENT RECLAMATION PRACTICES

Land reclamation has been occurring in some fashion for many centuries. During recent decades Earth healing has taken many paths, some good and some not so good. Americans have seen their land as boundless and have ruined it for four centuries by pulling up stakes and moving on, generally in a westward direction. Resident Europeans, with a longer history of land care and limitations on arable land, recognized that their land needed healing in a rational manner very soon after disturbance. For a century and a half in western Germany, surface coal operations have proceeded in an orderly manner by taking topsoil (in recent years on a conveyor belt) to the previous day's site and replacing the just-removed topsoil. Valuable soil bacteria are hardly disturbed. However, in many American rural and urban development projects, topsoil is buried, mixed with subsoil, or piled and stored for months, during which the soil is essentially sterilized.

Time is of the essence in erosion control, while profits demand quick solutions utilizing heavy machinery. After all, the land is meant to yield coal, and nature will heal it in due course—even if this is years and years away. Can this Appalachian chapter of the tragedy of the commons be averted? There are dollars to be made, so the regulations regarding buffer dams and returning the land to its original contours are obeyed only to the degree that the bond will be released. Subsoil is exposed along with acidic shale. Irresponsible companies spray green dye on the surface, scatter a little straw, and sow some grass seed, leaving the rest to chance. Often the intended ground cover does not survive. As a result, more responsible coal companies have moved beyond grass, which has failed to provide an adequate cover, and native tree plantings, which are slow to grow and time-consuming to plant. Even the native plants that are briars and brambles are cynically considered to be an established species on the wasted land—but wild berry cooperatives, which could make productive use of reclaimed land, have a difficult time making ends meet. A little green cover may be too little, too late.

Returning land to predisturbance conditions demands that the landscape be covered with good soil, not subsoil. Furthermore, current federal regulations require that there be no gullies or major water erosion, and even spell out details about check dams and rock-lined drainage systems. Without the proper topsoil mixture, revegetation options are limited. Those options expand rapidly after proper treatment, such as liming to remove acid-induced coal by-products. However, regrowing the vegetative cover that existed before the mining operation is nearly impossible, at least in our lifetime. Returning to original vegetation is a dream that is usually overcome by the reality of new shopping malls or parking lots carved out of the disturbed land.

Generally the practice in America in richer soil areas and on moderately sloped land is to reclaim land in grasses (not necessarily native) and use the land for pasturing cattle and sheep. The Eastern Ohio Resource Development Center, which was established in 1965 with 728 acres in Noble County and has since been given an additional 1,325 acres of strip-mined land, has tried to develop improved pasture for year-round livestock production. The project has harvested hay to be fed to cattle and sheep and has also cleared land, fertilized fields, and constructed ponds and fences. Current research involves planting available forage crops to provide minimum input for maximum return.

FUTURE RECLAMATION

A host of ideas spring to mind when one sees unreclaimed land. Some of these are more practical and efficient than others.

Tree Farming

Some effort has gone into replanting native species in a regenerating forest in much the same way that species return to land disturbed by fire, flood, or tornado. Thus attention is first given to succession species, precursors to the climax forests of tulip poplars, oaks, hickories, and chestnuts that will follow in due time. Black locust and some of the aggressive pines are found to tolerate and even thrive on depleted and acidic soil on coal-mined reclamation sites. Reclamation involves costs, and harvesting these succession species for wood pulp and wood chips can provide revenue while leaving slash, which helps restore biomass for future nourishment of depleted soils.

Some have suggested that the raising of Christmas evergreens is another possible moneymaking practice on reclaimed land. But such lands take preparation and liming, and growing ornamental evergreens can only be a successful venture if part of the land preparation is done prior to turning the land over to private tree growers. If evergreen growers are required to bear liming and erosion-control expenses, the cost will be too high. A possible solution is for private tree growers to receive a land-restoration subsidy from government reclamation agencies or mining companies as part of a public-private partnership.

Grape Growing

Several times in the past few decades coal companies have suggested growing grapes on surface-mined land; after all, grapes from the time of the exodus from Egypt have been the sign of a promised land. Recently Coal-Mac Inc. announced plans for a vineyard and winery on 200 acres at the company's Phoenix 3 operation in Williamson, West Virginia, where loose soil and irrigation facilities are available.

For successful grape growing, some organic matter and nutrients must be added to the acid subsoil. Grapes can grow on rather poor soil, as has been shown on other continents. The quality of the grapes depends primarily on weather conditions; the overall climate, especially in late summer and autumn; the type of grape; and the topography. The land must be well drained, and the amount of average sunlight is highly important. Appalachia has always had wild grapes and is able to continue to grow them—even on reclaimed land. Grape juice and wine can be processed locally, creating more opportunities for local employment and tax revenue.

Other Uses

People have many ideas on what to do with unreclaimed land. Some of the plans are far-fetched, such as the use of lands for all-terrain vehicle tracks. This is occurring in McDowell County in southern West Virginia, and similar ideas have been proposed by local development officials for Harlan County, Kentucky. ATV drivers prefer muck and mud, and there's an abundant supply at unreclaimed surface mine sites. But ATV riders often seek undisturbed wooded areas and even pristine areas, so they gravitate to undisturbed neighboring woodlands. Today, land uses of choice for stripped land are industrial, commercial, educational, residential (including mobile home parks), and other forms of land use requiring blacktopping of vast parts of the stripped area.

Creating ponds and lakes had a certain appeal in the past, but runoff from disturbed land results in impounded water that is too acidic for fish production. Institutional use of land has had a reasonably good track record. West Virginia Jesuit University is built on a reclaimed mine site; after four decades in which considerable attention was given to building organic matter in the soil, its flowers and garden are growing with substantial vigor.

The hope presented in this section, however, is that damaged land be returned to forested conditions that existed prior to disturbance. That presents a considerable challenge to those convinced of the importance of modeling ecological restoration as part of the Earth-healing work to be done in Appalachia. Research is needed to determine how a wide variety of mixed mesophytic forest species can be employed in an order resembling the succession of natural vegetation.

CHAPTER 18

Retreat Cabin Sites

After 31 years of experimental trial and error, I am finally building my small isolated retreat cabin on an abandoned cleared farmhouse site; this new structure lacks for none of the amenities of modern life. With photovoltaic electricity, passive solar heating, heat storage, passive solar cooling, and my water supplied by rain, all current conveniences are mine without the usual charge. This has been done on my 41.75 acres of forestland in the most diverse, and in my opinion the most awesome, forest in the world. Couple this with the Red River Gorge geological areas . . . it's like, why have I been so favored?

Syl Yunker, Gathering Place:
A Permaculture Plantation,
Stanton, Kentucky

The need for rest and reflection is a basic component of the psychic and spiritual health of all who seek to be Earth-friendly. Those of us who see the importance of a wholesome environment for a good quality of life realize that we need an ecologically sound atmosphere where we can meditate, pray, come into balance with nature, and refresh our souls. We do not deny that some who are forced to sustain their spiritual balance in extreme poverty or in prisons can find peace of soul under extraordinary circumstances. Normally, though, people seek simple places to which to retreat for at least a short period.

This is not a luxury for the rich but a necessity for all. Native Americans, South Sea Islanders, Mongolians, Alpine inhabitants—people of every time and place—have sought remote retreats. This chapter is directed to those who seek the divine in scenic places devoid of pressures and discordant noises but full of fresh air and sunlight. Our hope is that these ideal places will influence folks to find time to use them.

APPALACHIA: AN IDEAL RETREAT

Formal retreat places are meant for some people but not all. The mountaineer free spirit is vocal in insisting that a retreat place is important for rest and proper reflection. A rigid, filled schedule does not appeal to such a spirit. And the need for reflection and rest is greater today than in the past because of the unprecedented mobility and stress in our superconnected, informed, and wired populace. In one way the fast-moving modern age is similar to the unsettled conditions of the late Roman Empire, when hermitages grew in popularity in the Egyptian and Syrian deserts and other portions of the Christian world. While that time was certainly a watershed in seeking solitude, throughout the ages and across cultures, solitude-seekers have sought woodlands, mountaintops, lakeshores, or any suitable place to put up a shelter from the elements. We need to see that such places are within the reach of average people and that they can be built and maintained at low cost.

Appalachia is blessed with many ideal locations for retreat cabins and is within a day's drive for about half of all Americans and virtually all easterners. Within these mountains are numerous coves and hollows, hillsides and scenic overlooks, creek banks and lakeshores where people can rest. The cabin does not have to be on the side of a highway or in a developed settlement or equipped with all the amenities. It can be simple, quiet, secluded, and affordable.

We focus here on the cabin builder/owner who wants a place away from others, though it may be used by other visitors on occasion. To a lesser degree, this chapter may be helpful to people with access to land that they don't own but on which they are allowed to establish a simple nonpermanent dwelling—not an affluent person's second home but a temporary retreat site. Hunters and fishers have their huts, youth have tree houses, and some people have caves and unique rocks where they go just to get away. These favorite spots are sacred spaces, whether in natural or artificial settings. A simple retreat should be within the means of most people who are already landholders. Furnishing such a cabin or place that provides shelter from the elements is an Earth-friendly gesture, because it is the most suitable way for the average person to get close to nature for some period of time. Thus we focus on building at a suitable place, and not on just finding a site and roughing it there.

We all need to rest, but we do so in different ways. One person's rest may be another's torment. Some people prefer a backpack retreat in the wilderness; others prefer to spend time in a rustic but somewhat protected setting. And the same person may have different preferences at different stages of life.

SELECTION OF PLACE

> *Indian campsites were judiciously chosen with many factors being weighed in the selection process. Safety and the availability of necessary resources were prime considerations. The scouting and evaluation process by seasoned members of the band usually yielded a location admirably suited to the needs of the occupants.*
>
> Helen Graham Bronson

The ideal Appalachian retreat cabin has certain characteristics, including accessibility, simplicity, tranquillity, and seclusion. Those and other attributes are discussed here.

Many users prefer a site that is easily accessible, and today this generally means by vehicle—perhaps even an electric golf cart charged by solar energy. In choosing a site, ask questions like the following: How close is the place to where the user resides (most potential users prefer places that can be reached with little motorized travel)? Is the place accessible by a major highway? Many retreatants want places that are not right on a major highway but that are not reachable by an overly rough or winding road. Would it be very difficult to bring in supplies to build the cabin, and later to bring in food and essentials for one's stay? A good access road to the site is a great advantage for older and partly or totally disabled people. Ideally, a retreat cabin would be near a road and yet not so accessible as to bring unwanted visitors and trespassers.

The retreat cabin should be inviting, clean, and still rather austere. The seeker may be willing to engage in a simpler life. However, there are degrees of simplicity. Some people—especially older folks or those with some disability—need conveniences such as a resting place that doesn't require a ladder to reach, an entry without too many steps, floor space on a single level, or good lighting and ample windows. While the general rule is the simpler the better, with a little extra effort, amenities can be available such as interior heat, a nearby compost toilet, or cistern water. Simplicity has an inherent value and should not be defined solely by remoteness from building materials, and the creation of a luxury apartment in the woods is not a worthwhile goal.

The site ideally is quiet. While some can make do with a "silence of the heart," normally people prefer external tranquillity. Generally, silence and seclusion take precedence over scenic views. If the site is near a road, traffic noises can be counteracted by establishing vegetative, wooden, or masonry noise barriers, or by selecting a site where the topography—valleys or coves—deflects sound. A scenic view is an added benefit, whether of a grove of trees, a mountain vista, a shimmering lake, or a waterfall. Can the cabin dweller enjoy a beautiful sunrise or sunset? Can he or she find fellowship with other creatures, attracting wildlife by providing bird feeders or salt blocks for deer?

People on retreat like to be separated from others, secluded but not isolated. They prefer not to feel totally alone or vulnerable and want to be in communicating range of others without mingling with them. If the cabin is in a well-chosen enclosed natural space (a cove, valley, hillside setting in dense woods, etc.), the personal need for seclusion outweighs the need for interpersonal connections. Most solitude seekers prefer relative seclusion, that is, a certain amount of privacy while still being able to communicate with other people. If two or more cabins are contemplated on the site, place them a short distance apart in such a way that the entrances and patios are private. The ingenious placement of the cabins at St. Catherine's at Springfield, Kentucky, allows for both physical proximity and a sense of seclusion. This is done by rotating the buildings so that the front of one is not visible from the other. For some isolated cabin dwellers, a radio, telephone, or a pet animal may satisfy the need for contact. Urbanites and those more vulnerable because of age, physical condition, or gender may have different perceptions on this subject.

Some environmentalists may object to the development of retreat cabins in secluded areas, but the areas we are talking about are not remote wilderness areas—for one thing, they are on existing access roads. While we have promoted retreat cabins to over sixty different institutions in the last quarter century, none of these are in remote or primitive wilderness areas. The Gathering Place mentioned in the epigraph to this chapter is in a clearing in the forest where an old farmhouse once stood.

Retreatants desire contact with flora and fauna in natural settings like forests, isolated islands, mountaintops, or rock overhangs. Ideally, the cabin is shady in summer and has good southern exposure for winter, spring, and fall; it should be protected from winter winds and open to fresh air and adequate breezes in the warmer seasons. Trees add intrinsic value to the experience and may also provide fruit and nuts. An eremitical environment was described by Saint Manchan of Offaly fifteen centuries ago:

A pleasant woodland all about
To shield it [the hut] from the wind,
And make a home for singing birds
Before it and behind.
A southern aspect for the heat
A stream along its foot,
A smooth green lawn with rich top soil
Propitious to all fruit.[1]

Just as a house may or may not be a home, depending on the love involved in building, decorating, and maintaining it, so the same can be said for a retreat cabin. The cabin should have an air of hominess and hospitality, so that the owner can share it with others, either while the owner is present or separately. Such sharing may mean being prepared to forgo complete privacy and share cabinets and living or resting space.

It is good stewardship to consider that the cabin will be transferred to others in due time, so salability is a consideration. (Many regard retreat cabins as a growth industry, given the desire of growing numbers of urbanites and suburbanites for a quiet place to rest.) Independent road access, future development and scenic preservation, pleasing appearance, and durability of the structure should be taken into account when starting to site, design, and build the cabin.

The structure should not be overly prominent and thus mar the landscape; in fact, the cabin builder should not change the landscape in any noticeable fashion. Sometimes to avoid this it is necessary to take a cue from the hermits of old and move to caves or other underground sites to reflect or live, perhaps by building an underground or semi-underground house (see chapter 21). In other cases, a secluded cordwood or cob house or other building that will not disturb the scenery in any significant way is appropriate.

1. Thomas Cahill, *How the Irish Saved Civilization* (New York: Doubleday, 1995), 152–53.

CONSTRUCTION SUGGESTIONS

The Earth-friendly cabin builder should think of small shelters built with native materials in natural settings. No one wants to spend a lot of time maintaining a cabin intended for rest. In fact, a larger building may feel rather lonely, while a smaller building that blends with the natural surroundings is more comforting. Good architects and builders design a structure taking into account the unique features of the site, as was done in the Long Branch orchard house shown in figure 18.1. This is especially important in highly scenic and fragile terrain, where effort is required to preserve the area's ecology and beauty. Sometimes misguided owners and builders look out away from their structure to distant unspoiled views, not taking into account the effect their building will have on others' vistas. And building on mountaintops is not only detrimental to the landscape but also dangerous because of lightning strikes.

An added advantage of simplicity is that small dwellings cost less, are good ecological educational models, and afford occupants an opportunity to practice self-reliance and dependence on God's creation for nourishment. Yurts and cordwood buildings are good examples of low-cost shelters using native materials. Other designs such as domes and subterranean structures are also possible. The point is that the structures need to be low-cost, nature-friendly, and suitable for prayer and reflection.

Many people see a beautiful location, fall in love with it at first sight, and decide to build there. But it is better to go slowly, following the example of hunters and naturalists, who often move about and camp in several locations before choosing a site. For prospective cabin builders, it is better to establish a temporary site such as a tent platform and spend some time getting to know the site. Is this the perfect spot, the sacred place? Final site choice should result from careful reflection. Only when the right site is determined should a more permanent structure be built.

Fig. 18.1. The orchard house at Long Branch is a retreat cabin.

The cabin or other structure should be made of native, nontoxic materials. Throughout history people built with local materials (rock, rammed earth, cordwood, tree waste materials, etc.). Only in the resource-wasteful times in which we live are building materials transported over great distances. Many, but not all, appropriate technologists consider a 500-mile radius acceptable. It may also be possible to build a small traditional house using locally recycled materials, which are often found in abundance.

New materials should be low in volatile organic compounds (VOCs) such as formaldehyde; older salvaged materials should not contain lead-based paints or other toxic materials no longer allowed in building materials. Quite often plywood and rugs have adhesives and other compounds that can contaminate the indoor atmosphere. For instance, different brands of liquid nails may differ tenfold in VOC content; plywood differs dramatically in types and amounts of adhesives, so care in choosing is necessary.

A good rule is to build small but comfortable. What this mean depends upon whether the structure is for year-round use or only for the milder seasons. Incorporate designs that will keep the place small, cozy, and well adapted to residents' needs. The square footage needed may be minimized by building a sleeping loft for the able-bodied above a single-room living, reading, cooking, and dining area. Beds take up space; instead use a sofa bed or futon or bedrolls on the floor. Low-ceilinged areas in the loft may be roomy enough for a chest of drawers, a reading lamp, and a bookshelf. For proximity to nature, equip the exterior with a shady porch and shade trees for natural summer cooling.

For privacy and thermal barriers, use plants, not fences, whether of wood, concrete, or metal. Vegetative materials are more cheerful and will blend in with the countryside. Remember to select native species, not exotics, which can grow out of control, to the detriment of native plants. For example, the multiflora rose was brought to this country to serve as a vegetative barrier or fence but spread throughout the countryside. Evergreens such as cedar, various pines, or decorative noninvasive bushes can furnish a thick and inexpensive natural barrier. These plants also serve as winter wind barriers. Plots of herbs, berries, and flowers could decorate the place, if there is time to care for them.

In equipping the retreat cabin, remember that it is not a motel. Air conditioning, television, wall-to-wall carpeting, oversized beds, room service, and ample showering facilities are not necessary or desirable. Instead, the cabin can be furnished with simple-lifestyle devices such as fixtures with compact fluorescent bulbs, a composting toilet, and a shower equipped with a solar water heater or a solar shower bag. Consider a cistern for a water source (see chapter 29). Gray water can be directed into an artificial wetland, which can serve as a flower garden. All biodegradable kitchen scraps should be composted in a designated area and some space allowed to sort all waste materials. The ideal cabin would have solar photovoltaic arrays for generating electricity, solar path lighting, and passive solar space heating. Earth berming or superinsulation of the sides and roof could offer major long-term savings. No air conditioning is needed if the place is airy and shaded. Window shades can be installed to keep out the sun, and quilted window shades will conserve heat during the cold season.

USING SACRED SPACE

Sacred sites are those locations where the heart, mind, and whole being are moved to the wonder of Creation and the Creator. Most settled communities with a spiritual heritage have found and identified their own "sacred sites." These may have cultural or historical significance, such as the site of a great event or a traditional communal gathering place. However, there are also sites that are sacred to individuals or couples, a spot where they come to watch the sunset, pray, or just commune with nature.

A site that stimulates the various senses—the beauty of a scene, the scent of evergreens or seawater; the sounds of wind, birds, or rushing water; the texture of rock or tree bark; and the taste of sassafras or nearby berries—is a natural meditation area. Typically, sacred sites are rather secluded and quiet, so that they are conducive to prayer and reflection; they should not be inhabited or built upon. In some cases the sites are plainly visible and known to all; others are held in secret by a few for fear that they will be too frequented or discovered and trashed by trespassers.

The owner of a retreat cabin and visitors to the cabin have an opportunity to find their own sacred spaces to rest and reflect.

In selecting a site, some attributes to look for include seclusion, accessibility, naturalness, lack of major distractions such as noise, and conduciveness to prayer and reflection. As with the retreat cabin, the sacred site is meant for nonresidential visits on particular occasions. Again seclusion but not total seclusion, accessibility but not total ease of access, and a peaceful atmosphere are necessary. Sacred sites are not necessarily ones that should be inhabited or built in; in fact they should be left in their natural condition because of their unique service to seekers of rest and reflection.

CHAPTER 19

Energy-Efficient Passive Solar Design

The sun is an integral part of my life for the past two decades. My solar home has lighting powered by PV panels on the roof, and the building is designed to furnish solar warmth in winter. The south-side glazing gives the place a feeling of being free and airy. The sun also helps grow the plants in the greenhouse on a lower floor. A compost toilet suffices for handling waste and a cistern satisfies the water needs of this residence. Shade trees outside help keep the place cool in summer. I can't ask for anything more as far as housing goes, for the place inspires me in my design work.

Mark Spencer, coordinator, ASPI solar house

This chapter deals specifically with solar shelters; others deal with solar-generated electricity, space/water heating, and gardening. Low-cost residential shelters can be built by individuals who want to be free of mortgage and energy bills while enjoying comfort, coziness, and environmental harmony. Solarized structures in the temperate zones, whether residences or institutional buildings, provide warmth in winter and natural cooling in summer. The better the building is situated, constructed, glazed, designed, and fitted with thermal mass for heat retention, ventilation, and insulation, the more comfortable, low-cost, and easily maintained it will be.

In the Northern Hemisphere, the ideal solar house is built facing south, with its rear protected against harsh north winds by earthen berms; it is constructed of durable native materials and designed with properly placed rooms in the sunny portions. It has sufficient glazed areas providing daylight through heat-rejecting sloped glass covering the sun space. It has sufficient internal thermal mass to keep the rooms warm in winter and cool in summer. It has access to fresh air and the ability to recirculate heated air to all parts of the structure. It is superinsulated with long-lasting, vermin-proof materials. Additional features may be air locks at frequently used entrances, geothermal heating (though heat pumps require energy from utilities), photovoltaic arrays on roof or ground, and daylighting with properly placed windows and occasional skylights.

SOLAR SITING

Proper siting is a key to success in designing a solar house. An ideal solar house in the Northern Hemisphere would be built into a south-facing hillside so as to maximize the sheltering and insulating effects of the earth. An educated guess could approximate the annual shading at given locations, but this can be shown as well with a simple solar pathfinder like the one at www.solarpathfinder.com. This device shows the amount of time solar rays strike a given surface (insolation) throughout the year. Buildings should face south or even a little east or west of south with no major shading.

In some cases, it is necessary to remove trees where solar has been installed, as was the case at the ASPI solar house, shown in figure 19.1. There, fast-growing poplars have had to be cut after only two decades to allow sunlight to be maximized in winter. Often the trade-off is between shade in summer and sunlight in winter, and some people prefer more shade. For attached solar greenhouses, maximum sunlight is required for winter horticulture, so more winter sun is preferred to more summer shade.

Fig. 19.1. ASPI solar house

SOLAR CONSTRUCTION

The cliff dwellers in the Southwest established south-facing dwellings so that the lower-arching sun of winter sent solar rays deep into the cliff residences when warmth was needed in colder weather. On the other hand, in summer when the path of the sun is higher in the sky, the cliffs were shaded and the dwellings remained cool. Passive solar buildings are constructed in the same way, with south-facing windows to catch the lower winter sun's rays in winter and eaves to keep the summer sun from causing an overheating problem. Balconies and other protruding constructions help provide shade from the summer sun. Such passive solar design elements are less expensive than using active systems, which employ mechanical means to move solar energy.

Many ideal passive solar house features are found in the main Long Branch solar structure known as a southern

Appalachian bioshelter. Its attached solar greenhouse faces south and the building is set into a wooded, south-facing cove to protect against winter winds. The house is a multistoried construction complete with a two-story greenhouse and passive solar compost toilet, office/living space on the main floor, and office/bedrooms on the upper floor. The building has double-glazed windows on all floors and additional thermal mass in its fully insulated (2 inches of rigid foam around the perimeter and beneath the slab) concrete slab-on-grade floor system. Rot-resistant shelving for plants, made of cement board painted dark blue, also serves as thermal storage for the greenhouse, as do several hundred gallons of water-storage containers encased in the poured-in-place concrete steps.

The wall system is 2-by-6 frame construction built with locally sawn tulip poplar wood. It has 1 inch of rigid foam exterior sheathing and 6 inches of fiberglass insulation for an R value of close to R-30. Ten inches of fiberglass insulation is used in the ceilings. Careful attention was paid to sealing floor plates and around all doors and windows to prevent any air infiltration. A naturally reflective galvanized steel roof was used to reflect unwanted summer heat. Deciduous tree shade is found on the east, south, and west sides of the buildings to facilitate natural cooling.

The ASPI solar house is located on a steep slope overlooking the Rockcastle River. It was built with minimal land disturbance on a minable coal seam in an Appalachian pole-house fashion. The construction was directed by Jerry Nichols, with the help of two dozen volunteers over a four-month period. The 1,750-square-foot building, which cost $20,000 when it was built in 1979, has a number of environmentally friendly features: a south-to-southwest orientation (surrounding hills cut into the total insolation time), a Clivus Multrum composting toilet, an attached solar greenhouse with a 1,000-gallon fish tank for heat storage, rock (formerly water) thermal mass on both first and second floors, a photovoltaic system for lighting, a backup wood-burning stove, an ample 4,000-gallon cistern supplying rainwater, terraced wetlands, a root cellar behind the greenhouse, decking on the south, east, and west sides, an abundance of oak and poplar shade trees, which reduce traffic noise on the west side, and indoor plants for air freshening.

SOLAR GLAZING

Passive solar structures have relatively large window surface areas on the south, moderate amounts on the east and west, and scarcely any on north walls. Windbreaks are present on the side more subject to winter winds. Additional protection includes wall insulation and double-paned windows. Windows can be opened and closed to provide cool air in summer.

The ASPI front window space was originally equipped with plastic sheets called Kalwall, which became clouded after a decade even after a recommended chemical treatment. These were replaced by double-glazed glass panels, and the chemically treated Kalwall was found to be highly flammable, a feature never mentioned in the literature. We therefore recommend against its use and counsel removing it where it is present.

Windows can be treated with glazing or films that can reflect at least 55 percent of the heat striking the window and over 90 percent of the ultraviolet radiation. Low-sensitivity (low-E) films can reduce glare by 60 percent. At a cost of about 75 cents a square foot, payback occurs in twelve to eighteen months. Low-E window film costs a fraction of "factory high-tech" Low-E windows, which have a payback period as long as thirty-eight years. The life of the film is about fifteen years, and applying it has the added benefits of making the glass shatter resistant and minimizing ultraviolet light exposure, reducing the fading of rugs and upholstery. The film provides high light transmission, and it can be used on most glass.

THERMAL MASS: HEAT RETENTION SYSTEMS

Water has a high specific-heat content and the capacity to retain and release sizable amounts of heat in cooler periods. Fish tanks, barrels, gallon jugs, and columns are among possible receptacles for water for heat storage. Some designers are quite creative in integrating water-storage containers into the solar structure. ASPI has water tanks in its greenhouses but removed a series of 55-gallon barrels for storage on two floors of the ASPI solar house when these started to leak after two decades of use. In place of water, a rock thermal mass was installed on the floors, replacing tons of water with an equal weight of rock in this cantilevered structure. However, heat storage in water is two and a half times greater than with rock.

Some solar builders gain thermal mass through tile floors, which absorb heat and radiate it out during cool winter nights. Such flooring takes up no extra space, does not interfere with furniture, keeps the house warm, and even provides a consistent, elegant feeling. Others use masonry in the form of freestanding or attached walls to gain thermal mass. Back walls are often made of heavy masonry and painted a dark color so that when reached by the winter sun's rays, they absorb heat energy and release it at night. The walls may also be used for decoration and other purposes.

VENTILATION

Water or masonry thermal mass releases its stored heat energy at cooler times through radiation. This could be augmented through mechanical means such as electric fans. However, when well integrated into the basic construction with a series of air ducts and vents, the system will allow natural thermocirculation, with air passing across the thermal mass surfaces and being warmed. Through convection, the warmed air can be carried to other parts of the structure's interior. Such ventilation systems are standard in a wide variety of modern homes.

Proper ventilation and all forms of heat transfer demand proper design and planning before construction starts. Many resources are available: books, publications, Web sites, consultants, architects, and information gathered at solar home tours. Aspiring solar home builders will not become instant experts, but they can learn to help provide for a proper flow of air in their homes so that they are comfortable in summer and winter. Summer comfort also depends on exterior shading (see chapter 6) and on other natural cooling methods and practices (see chapter 20).

INSULATION AND ENERGY CONSERVATION

Ideal passive solar structures emphasize superinsulation, for solar energy gained should not be lost. Let's review insulation in greater detail for all aspects: materials, weather stripping, and caulking.

Over twenty-five years ago, it was estimated that 30 percent of the then 80 million American residences were not insulated at all.[1] While many of these homes have since been retrofitted with insulation or torn down, approximately one-tenth of American homes were not insulated in 2005. In one environmental resource assessment, I confronted a southern institution manager with the fact that a window was open on a cold January day. He assured me the place had funds to pay its energy bills. I returned to give a lecture six years later in the same assembly hall and found the window still ajar. I assumed their finances were holding up—but the environment had suffered. Economic wherewithal cannot substitute for respect for the planet's health and well-being.

---

1. Robert Stobaugh and Daniel Yergin, eds. *Energy Future* (New York: Ballantine, 1979), 212.

Air leaks caused by wind or pressure differences inflate heating bills in winter and cooling bills in summer. Leakage can also occur at soleplates, wall outlets, external doors and windows, fireplaces, and kitchen and bath vents. Energy-conscious people know that insulation has a rapid payback and that conservation is always the first priority, even taking precedence over passive solar building.

Consumers determine their insulation needs based on their heating zone location and the R value (measure of resistance of insulation to heat flow) of their materials. These values are listed on every commercial insulation package. In shopping, consider several factors: price, ease of application, the area needing insulation, and availability.

Some rock wool, glass fiber, and cellulose fiber must be blown into spaces with special equipment by a professional contractor. This is the method of choice for retrofitting walls and some ceiling spaces, but loose insulation is hard to handle and keep fluffy. Ceilings with no attic floor can also be insulated by installing batts, reflective foil side down toward the heated space, for a radiant barrier effect between living space and attic insulation. Cellulose insulation can be made by chopping newsprint and applying boric acid as a fire retardant. When buying cellulose insulated materials, check for third-party testing such as Underwriters Laboratory for fire safety and lack of corrosiveness.

A low-cost way to save heated or cooled air is through weather stripping, using commercial metal strips, wood, or adhesive-backed foam rubber, rolled vinyl with aluminum channel backing, rubber or neoprene strips, or felt strips (cheap but not very durable). Local hardware or home-building supply dealers can help in the choice. Solar building operators should give special attention to weather-stripping spaces between door and window frames.

Caulking is a good, low-cost way to winterize homes or offices with attention to foundation sills, corners formed by siding, entry points of outside water spigots and electrical outlets, wire and pipe penetration of ceilings, spaces between porches and main parts of a house, places where a chimney or masonry piece meets with siding, and places where walls meet the eaves at the gable ends of buildings. Caulking comes in the form of cartridges, fillers, ropes, and compounds. It is best applied in warmer weather and on clean surfaces. Plastic spouts on caulk tubes should be cut at an angle to allow for better "bead" control in application.

Oil-based caulking materials are the least expensive but also the least durable. In contrast, more costly polysulfide, polyurethane, and silicone caulk will last for several decades. Fillers made from hemp treated with tar, glass fiber, caulking cotton, or sponge rubber are used for larger cracks. Cartridge caulking is then applied. Rope caulking is good for temporary jobs around storm windows or air conditioners and can be reused for several seasons. In winter, seldom-used doors or large windows can be fitted with thick, rigid foam insulation panels bordered with duct tape to reduce wear. These panels can be installed in rooms during colder weather and may become banner energy conservers.

Early Appalachian settlers conserved in numerous ways. They even cut up their used garments for patches and quilts. The designs of their warm winter bedding are so artistic that many quilts are hung today as wall decorations. A way to combine art with conservation is to hang quilts in windows, which are major sources of heat loss. The production of quilted window shades, in fact, has the potential to be a profitable cottage industry, although quilting is very labor intensive.

Various types of less elaborate insulated window shades have been developed by a number of people, including Phyllis Fitzgerald of Louisville, Kentucky. One such shade uses materials of various types laminated with Astrolon, an insulating fabric and reflective material. The material is filled with batting and the outer and inner covers of the shade are sewed together in a professional manner. Rollers and attachments are necessary. Velcro may be used to hold the edges to the window frame. Another device in use at ASPI's nature center is 2-inch interior wooden shutters hinged to the long sides of the windows. These are opened to let down a rolled shade and then closed over the edges of the shade fabric to seal the edges so that no circulating air can create heat loss.

The passive solar house exemplifies Earth-healing practice and living in harmony with our environment. Americans pay substantial fuel bills for heating and cooling, bills that increased dramatically in the winter of 2005–2006. Owners of passive solar homes are insulated to some extent from these soaring costs—an additional incentive for promoting passive solar techniques. In place of a furnace fired by natural gas or fuel oil, a small backup woodstove for use during extended cloudy periods is all that is needed to maintain the interior comfort of a passive solar house.

CHAPTER 20

Natural Cooling

You may be surprised to learn that passive solar design strategies apply to home cooling as much as home heating. These strategies include proper building orientation, so that the house receives the sun when you want it (in the winter) and keeps it out when you don't (in the summer); the use of overhangs and shading to control the entrance of sunlight into the house; proper window placement and selection of appropriate windows to limit unwanted solar heat gain; adequate insulation, especially in attics; and the use of radiant heat barriers and ventilation in attics. The simple daily practice of opening up the house at night to allow the cool night air inside, and then closing it up mid-morning before the day gets hot, is another natural cooling strategy.

Andy McDonald, coordinator,
Kentucky Solar Partnership, Frankfort

Before the advent of mechanical air-conditioning, natural cooling techniques were used in residential construction as a matter of course in Appalachia. Higher elevations of the region were cooler and became a favored destination during hot, muggy summer seasons. Virtually no residence at higher elevations with some natural shading requires mechanical cooling at any time of the year.

The 2003 heat wave in France killed over ten thousand mostly older citizens. Temperatures were extraordinary for western Europe; thus many Europeans are now considering air-conditioning motels and homes. American domestic cooling is a major modern energy expenditure and results in summer electric demand spikes that can result in overloads and blackouts. Cooling and heating use more energy than any other domestic application. Air conditioners alone use up to one-sixth of U.S. electricity and, on hot summer days, consume 43 percent of the U.S. peak power load. They cost $10 billion a year to operate. According to the U.S. Department of Energy, heating and cooling systems in the United States emit over a half billion tons of carbon dioxide into the atmosphere each year, adding to global warming. They also generate about 24 percent of the nation's emissions of sulfur dioxide, a chief ingredient in acid rain. In 1960, about 12 percent of American homes had air conditioners, but that had risen to 64 percent by 1990. Today, most homes are equipped with air-conditioning units and people lacking these are regarded as living downright heroic lives.

In parts of Appalachia and elsewhere, underground caves, a constant cool temperature all year around, have been used from prehistoric days to modern times as hideaways or to cool food in the summer. In fact, springhouses and caves went hand in hand. Owners of nineteenth-century American farmsteads without nearby caves built half-submerged springhouses near natural seeps to store buttermilk, clabber for cottage cheese, baked hams, pies, meal leftovers, and ripe fruit and vegetables. These springhouses maintained a constant temperature of 55 degrees Fahrenheit—the natural refrigerators of that day. The cave and springhouse used natural coolness to the advantage of the homesteading family.

In this temperate zone, it has always been a challenge to build houses that are warm in winter but cool in the summer. In the late eighteenth and early nineteenth century, the preference was for houses built for summer living, since Appalachia had become a mecca for escapees from the river valleys and the Piedmont. In the Buffalo Trace region of Kentucky, where I was raised, there are many large, abandoned brick mansions built between 1790 and 1820. While well built for their day and of historic value, they were built as cool houses for hot, sultry summers. As a youth, I visited my great-aunt and great-uncle in their family place in Fleming County, Kentucky, built in 1812. I remember how cool it was in summer; the thermal mass of the large structure with its tall, airy windows and high ceilings, along with surrounding shade trees, made it a summer dream. Winters were another matter, with the house's small fireplaces barely heating the space immediately in front of them. That's precisely why many of these brick farmsteads were abandoned—they took too much renovation to insulate and heat in winter, for these were built for summer, and old-timers expected everyone to wear heavy woolen clothes in winter.

My family's farmhouse was built in 1931. It was eventually insulated and a central heating system was installed. Winters were always comfortable but summers were another matter. It was a hotbox for many years while shade trees were growing. As kids, we would escape to the cool orchard or to the cellar, which provided a place to play cards, make ice cream, and get away from the 90- to 100-degree summer heat. The cellar was also where the preserved food was kept. Our cellars taught a lesson: one can go underground for combination summer and winter shelters and do so with no additional energy use.

The use of air-conditioning, even with an integrated heat pump, requires the expenditure of energy that generally comes from nonrenewable sources. The monetary cost of this energy as well as the labor and equipment required to install an air-conditioning system makes home cooling a major domestic expense at the present time. The environmental cost of now-banned CFC (chlorofluorocarbon) refrigerants is observed in the ozone depletion of the upper atmosphere.

On the other hand, natural cooling, as demonstrated in the Long Branch pond cabin shown in figure 20.1, conserves energy and saves money. It can prevent a buildup of heat inside a home from solar gain through shade vegetation, awnings, window coatings, and window shades; prevent heat from entering residential space through earth-integrated techniques, insulation, constructed overhangs, and other methods; move air through the building

Fig. 20.1. The Long Branch pond cabin is naturally cooled, saving energy and money.

to remove excess heat through ventilation; and reduce interior heating sources.

For maximum comfort one should strive for a combination of techniques and not be limited to a single way of keeping domestic space comfortable.

COOLING BY BLOCKING THE SUN

Full use of sun-blocking techniques can save about 40 percent of the costs of cooling a building, including air-conditioning.

Vegetation

Vegetation is the first line of defense. In the chapter on energy efficiency (chapter 6) we mentioned the effects of deciduous shade trees situated on the south and west sides of a building. Vegetation becomes the structure's straw hat, reducing the temperature of roofs, walls, and driveways. Deciduous trees provide shade in summer and allow radiant energy to pass through in winter; they can save between $100 and $250 per year on energy bills. Also, trees pump water from the ground into their leaves, and as this evaporates, the trees and surrounding area are cooled. Trellises covered with native vines also provide shade and evaporative cooling. They can also shelter air conditioners, heat pumps, and evaporative coolers and thus improve the performance of cooling equipment. Fast-growing deciduous vines such as grape and Virginia creeper are attractive design solutions that lose their leaves in winter, allowing the sun's heat to strike the building. Vines should be on the hottest sides of the house, usually the south and west, and a small distance from the wall to protect it and offer a buffer zone of cool air. Shrubs may offer less shading than trees, but they cost less, do not have branches that can damage a building when broken, quickly mature, and take up less space. Shrubs can protect walls and windows from direct sunlight and yet allow maximum insolation for roof-mounted solar panels.

Shade Screens

For many years people have erected tenting devices (sunscreens, shade cloths, or solar shields) to help keep out the hot summer rays. Users have found that it is best to put these only on windows exposed to direct sunlight. Plastic or aluminum shade screens are lightweight and are not designed to keep out insects, only the sun—between 50 and 90 percent of the energy striking the outside of the window. While one can see through a shade screen, the view is somewhat obscured.

Awnings

A more conventional solar protective device (like the visor on a cap) is an awning that blocks high-angle sunlight in summer. Awnings can cover individual windows or sections of outside walls. They are more effective on the south and west sides of buildings. They come in a variety of sizes, shapes, and colors. Some awnings are made to stay in a fixed position while others can be rolled up in winter or during cooler parts of the day to allow low-angle sunlight to reach the building. Before the advent of air-conditioning these were used on most American businesses to protect merchandise from the sun and keep the place cool. They have the disadvantage, however, of blocking the view from the top half of the window.

Window Coatings

Modern window manufacturers offer heat-reflecting coatings that block out up to 25 percent of heat gain without affecting the penetration of natural light. Low-emissivity (low-E) coatings cut heat loss from the windows in the winter. The coatings can be designed for primary use in either summer or winter. Coated windows can be custom-made to take into account the direction of the various windows in the house. Still, in spite of coated windows' insulating properties, people tend to prefer uncoated windows because they offer better visibility.

Window Shades

What shade screens are to the exterior, window shades are to the interior. They include roller shades, venetian blinds, and drapes, especially double ones, which conserve heat in winter but can keep a room cool in summer by reducing heat gain. But interior shades are generally less effective than exterior ones. Interior shades work in three ways: reflecting light back out the window before it is converted to heat, blocking hot air movement around a room from the window area, and insulating hot surfaces of the window glass. These shades act best when they are light colored or reflective on the window side, fit tightly to prevent air movement, and are made of insulating material (see chapter 19).

Underground and Earth-Integrated Houses

Underground buildings are snug in winter and cool in summer, and they can be built at reasonable cost, but an effort must be made to control excess interior moisture in Appalachia. Earth-integrated structures have a way of moderating great temperature differentials in different seasons.

Some people do not like to live underground, for they want a less closed-in condition. The partial earth-integrated house using passive solar design allows for better contact with the immediate exterior environment. This is done by allowing the south-facing external wall to be open, maximizing heat gain in winter. Earthen berms used on external wall surfaces provide insulation in winter and cooling in the summer. All types of earth-integrated houses benefit from the relatively constant earth temperatures (see chapter 21).

Overhangs

Most homes have a built-in shading device in the form of an overhang that blocks high-angle summer sun and allows low-angle winter sunlight to enter. Fixed overhangs are always a compromise, since the spring and autumn sun is at the same medium angle. A roof overhang on a two-story house will not shade a first-story window, nor will a overhang provide relief from the overheating effects of west windows.

Insulation

Whether the air to be retained is heated or cooled, the same principles apply. Thus many of the techniques of passive solar heating fit well with natural cooling strategies. Insulation and weather stripping that prevent heat loss in the winter will also retard heat gain during summer. Insulating a ceiling will keep heat from penetrating into the house. A minimum of R-30 fiberglass insulation is relatively inexpensive and can save energy in both summer and winter.

Attic Vents

Convected airflow is increased by increasing the vent area. One sees rotating ventilators on roofs that allow for the convection of hot air out of the house. When these devices turn in breezes, the ventilation rate increases. Trombe walls that can be vented to the outside during cooling seasons are another built-in ventilation method. The sunlight hits the wall and heats air in a space between window and wall to temperatures above 150 degrees Fahrenheit, allowing for a convection of this hot air out of the house, at the same time that the cooler air coming from the north side of the house is drawn in. Passive solar houses work as well to remove hot air through a solar chimney effect.

Inexpensive gable roof vents on each end of the roof greatly reduce the amount of accumulated heat that otherwise would radiate into the interior. More-effective attic ventilation can be gained through the use of a ridge vent, which is quite conventional in recent construction. Commercially available vents are lightweight and durable and eliminate the need for turbines or louvered vents. They are also designed to keep out bugs and wind-driven rain. These vents can be installed by the homeowner on new roofs or as an easy retrofit to existing roofs.

Other Energy Savings

Movable shutters on the outside of the building can be used to retard summer heat gain as well as to keep heated air in houses in winter. In much the same way, thermal mass can act as a heat sponge by slowing air temperature increases during the day. The mass gives off some of that absorbed heat either by natural convection or by use of nighttime fans.

Dark-colored home exteriors absorb 70 percent to 90 percent of the radiant energy from the sun that strikes the home's surfaces. Some of this absorbed energy is then transferred into the structure through conduction, resulting in heat gain. In contrast, light-colored surfaces, especially roofing, reflect most of the heat away from the house. A radiant barrier or foil-faced paper stapled to roof rafters provides additional savings.

VENTILATION

The objective is to move cooler air through a building as much as possible throughout summer. Ancient tents were designed with spreading overhangs and open sides that allowed air movement in summer, thus cooling occupants. Using convection principles, many buildings are built in the tropics and semitropics to be naturally airy, with strong convection currents. Often the site has been selected so as to get mountain updrafts or sea breezes. Where breezes are plentiful, high vents or open windows on the side away from the breezes will allow the hot air to escape. Since warm air rises, vents located at high points in the interior allow warm air to escape while cooler outdoor air flows in through low vents to replace it. The coolest air is usually found on the north side, especially if this is shaded. The greater the height difference between the low (intake) and high (exhaust) vents, the faster the flow of natural convection and the more heat that can be exhausted.

Appalachian people who live where there are sizable differences between day and night temperatures in summer can follow some very effective energy-saving cooling practices. When a breeze flows, the lowest windows on the side from which the breeze is coming can be opened. Interior doors and the upstairs windows on the opposite side of the house from which the breeze is emanating should be opened. The warm air in the house will draw upwards and out the upper window, an effect called thermosiphoning. Cooler night air entering on the windward side will sweep out hot air accumulated throughout the building.

Cross ventilation may be effective during the moderate seasons and during evenings. High-pressure areas occur when wind strikes the building and "backs up." Low-pressure areas occur where wind velocities are high and eddies form on the downwind side of the house. For maximum cooling, air moves into a building through smaller openings in the high-pressure regions and exits at low-pressure areas. Inlets and outlets usually should not be placed directly opposite each other. A greater area will be cooled if the moving air has to change direction in the room. Creating more green space or planting more native deciduous tree species will shade paved areas (concrete patios or asphalt driveways, which absorb heat from the sun); this is often a very important way to reduce unwanted heat gain.

Low-priced and easily installed, ceiling fans use little electricity—less than one-tenth that of air-conditioning, or about $1.50 per month for fans versus $20 per month (in 2004 figures) for an air conditioner. Fans also are less polluting than air conditioners. Models are available with reverse rotation, which can be used in winter to pull warm air down from the ceiling. Ceiling fans have become increasingly popular in recent years.

REDUCING HEAT SOURCES

Heat generated from within the house can contribute significantly to the costs of cooling. The following are energy-conservation suggestions:

Turn off incandescent lights—or better yet, replace them all with compact fluorescents. Only 5 percent of the energy that goes into a typical incandescent bulb comes out as light; the remainder comes out as heat. Compact fluorescents give off 90 percent less heat, as well as using 75 percent less energy, than incandescent bulbs. Initially more expensive, they actually cost much less than incandescents in the long run because they use less energy and last longer.

Cook with a microwave, outdoor barbecue grill, or pressure cooker. The microwave generates almost no heat and is much more energy-efficient than a conventional stove or oven. The pressure cooker generates less interior heat with relatively low energy consumption. Barbecuing keeps the heat outside.

Seal off the laundry room and line-dry clothes. Washers and dryers generate large amounts of heat and humidity, and dryers are heavy energy users. If you must use them, do it in the morning or late evening when you can better tolerate the extra heat. Vent laundry room air to the outside of the house.

Insulate water heaters. Water heaters radiate heat that can be easily contained by insulation. You can purchase a water heater blanket for about $20 at hardware stores, or you can use faced fiberglass insulation and duct tape.

Seal ducts and close basement doors. Many houses with central heating have ducts that run through the attic and crawl space. If the seams in these ducts are leaky, especially in the attic, they can draw hot summer air into the house, creating more of a load for air conditioners. Minor duct repairs are easy to accomplish, usually involving folding or crimping the tin edges with pliers. Air ducts that lead to the basement should also be shut off, as this part of te house usually cools itself naturally.

Shut off the gas supply to fireplaces and heaters. The pilot light generates a considerable amount of heat and should be off during warm months. Relighting the pilot light in the fall is as easy as pushing a button on most units.

Raise the thermostat on the air conditioner 10 percent to 15 percent for a period during the day. This can be done when sleeping or when you leave the house for a time. A programmable thermostat costs between $30 and $50, depending on the features.

In summer cooking, make large batches and refrigerate them for future use, or consider more no-cook meals.

HEAT PUMPS AND GEOTHERMAL ENERGY

Root cellars in Appalachian terrain stay at a rather constant temperature of about 55 degrees Fahrenheit throughout the year. This has wider applications than just storing vegetables and fruit. The ground itself can be a source of heat in winter, and in the summer air can be cooled by passing it through the earth.

Heat Pumps

Heat pumps first came into wide use after World War II, and these highly efficient sources of heat remain very popular throughout the United States. In simple terms, they work like air conditioners. A central air-conditioning unit concentrates the cold in a set of coils inside the house and is connected to a set of hot coils outside the house. If we attach a valve and flip the air conditioner around so that the cold coils are on the outside and the hot set on the inside, we have an elementary heat pump, which moves heat to the inside in colder weather. It takes energy to perform the job even though the equipment is highly efficient in comparison to baseboard resistance heaters. Icing can occur in the operation of the heat pump; to avoid pumping cold air into the house, the heat pump activates electric strip heaters or burners to melt the ice. These heating devices are deactivated once the ice is melted and the unit returns to heating mode. Throughout the process either electricity or gas is being expended; thus, nonrenewable energy sources are typically being consumed. Woodstoves may be used for backup heating.

Geothermal Systems

Geothermal heating and cooling systems use the ground itself rather than the ambient air to provide heating, cooling, and even hot water for homes with heat pump mechanisms. These systems, four times more efficient than fossil-fueled furnaces, move heat that is already in the earth, although they require an apparatus for movement. The units last longer than traditional heating and cooling devices because they are housed entirely indoors.

Geothermal systems use thermal energy stored in the ground to heat and cool. A body of water or groundwater could also serve as the source, but this is generally not as efficient. Proponents talk about low maintenance because the units contain few mechanical components, which makes them less prone to failure; the ground loop has an expected life of over fifty years and requires no maintenance.

The advantages of these geothermal system ground source heat pumps in comparison to gas, oil, and other heating systems cannot be overstated. Gas and oil heaters require flames and flues, produce odors, and are susceptible to leaks and carbon monoxide formation. Heat pumps can also produce dramatic savings in energy costs. However, we must not forget that it takes electrical energy to run heat pumps. While granting that they are energy efficient, we must remember that they are still hooked to the umbilical cord of the traditional utility system and nonrenewable energy supply. Heat pumps and geothermal systems would be far more environmentally friendly if connected to photovoltaics, wind generators, or fuel cells.

CHAPTER 21

Native Building Materials

We wanted to build this ["Earthship" rammed earth] house to address things like global warming, forest issues, and even some social justice issues. We feel that there are things that are more important than money.

Stan and Pattie Jones,
Madison County, North Carolina

A basic appropriate technology principle is to construct with materials that are close at hand. Importing marble to build a Taj Mahal may be an activity of the superrich, but in some ways, modern American culture imitates this practice by choosing to use imported (nonlocal) plastic, siding, asphalt roofing, and particleboard. Is it any different for those who seek to be environmentally conscious but import building materials from distant places (e.g., straw bales in humid Appalachia from distant, drier regions)? Our efforts should be concentrated on obtaining materials from local sources and processors who use local materials, including wood (see chapters 22 and 23), sand, clay (baked, pressed, or rammed), gravel (crushed or natural), and stone. Cement plants within the region may use local products, as do glassmaking operations. Let's concentrate on native wood, stone, pressed earth blocks, and clay.

WOOD

One of the ironies of modern times is that small Appalachian sawmills that could turn out high-quality, rough-cut pine and other wood products for framing and rough siding are unable to jump through current regulatory hoops. Lumber must be processed, kiln dried, milled, and graded to exact specifications for housing and other buildings—things that some local sawmills cannot do. Thus, local sawmill operators are frozen out of a market they once had to themselves, namely, rural buildings, fencing, railroad ties, utility poles, and even housing materials. All wooden products that must be cut to size, dried, cured, and pressure-treated in a special way are beyond their processing capabilities. The multinational corporations have superseded the local wood processors, and thus communities suffer from loss of jobs and higher costs of imported materials—even in regions brimming over with forests and formerly prized for their high-grade forest products.

To make matters even worse, the effects of globalization are reaching into the forested areas of the eastern United States. While furniture factories close in North Carolina, the region's temperate forests, especially in pine regions of the Southeast, are being cut down limb, stock, and root, and the chips are being shipped to Korea and Japan, to return later as packing materials that find their way to stores within sight of where the chipping originally occurred. Furthermore, harvesting wood with heavy machinery is devastating our forests, and the practice of exporting raw products is devastating local labor markets. In addition, the returning product is higher priced and of a lower grade than what could have been made here from native materials. The local bioregion and the local economy suffer.

The chapter on cordwood buildings (chapter 22) tells about one area where local wood can be quickly used for constructing moderate-priced structures. Native wood was used by indigenous cultures prior to the European migrations and, once the Europeans arrived, was used for everything from log cabins and split-rail fencing to barn siding and bridges. Ancient American chestnut-sided barns still standing today testify to the durability of some of the wood materials used. Poplar logs covered with siding still undergird rural homesteads. Barn framing from red oak 12-by-12-inch timber can stand for a century or more provided the barn roofs are kept intact. Wood is available in many types, is easily worked, has enduring qualities, and has a natural beauty in its grain, color, and texture.

Wood Varieties

The following list is not exhaustive, but it shows the richness of Appalachian and eastern American woods:

Pine (white, Virginia, etc.) is the workhorse wood of the mountains and is much used in rough-cut siding, flooring, pressure-treated fence posts, bordering, and framing. Pine, like poplar, is lightweight and easier to handle than heavier woods like oak and chestnut.

Poplar is another traditional wood of Appalachia used for logs, siding, framing, rough flooring, statuary, and lower-priced furnishings and for fashioning a host of farm wood products. It is the wood of choice for utilitarian do-it-yourselfers.

Oak (white, red, etc.) is strong and can be used in a wide range of products such as framing, flooring, siding, fencing, implements, and furniture. It is tough and durable.

Red cedar has long-lasting rot-resistant qualities and is used for fence posts, decks, fine furnishings, roof and siding shakes, cordwood buildings, fine trimming, closets, and window framing, and where moisture might rot other woods, such as in a greenhouse.

Cherry wood is a beautiful red when polished and is highly prized for fine furniture and for mantels, cabinets, counters, furniture, and shelving.

Apple wood has many of the same qualities as cherry and is highly prized for interior decorative objects.

Maple is a beautiful wood that finishes well and is both easy to work and quite enduring when used for fine flooring and cabinet work.

Walnut (black) is a highly prized wood that is used in fine furniture, shelving, cabinets, and manufactured veneers. It finishes to a dark gloss that is favored by many wood lovers.

Hickory is a harder wood to handle and easily splits when dried. It is used for handles, smoking meat, and for certain specialty items.

Locust (black mainly) is long-lasting even in moist conditions, but it is difficult to handle or split. Locust poles cut to length are prized for fence posts and for rough barn framing, as well as other uses.

Much could be said about a host of other woods, such as the American chestnut, which will hopefully become favored again for its endurance when newly planted young trees become mature. Beech, ash, fir, and hemlock (which may be devastated by the imported Asian woolly adelgid) also are highly prized and beautiful woods for building materials. In fact, half the hundred-plus Appalachian and eastern American woody species have wood-product uses. The local bioregion and the local economy benefit whenever these woods can be cut, processed, and fabricated into products at or near home—as in the community center under construction in figure 21.1—rather than in distant places.

Woodworking

I grew up in a home where there had been a woodworking tradition going back to the Alsatian forests of my ancestors. My dad worked with wood from his youth and until his death carved and fabricated a variety of wood products, including spinning and wool wheels, ornamental devices, statues of famous Americans,

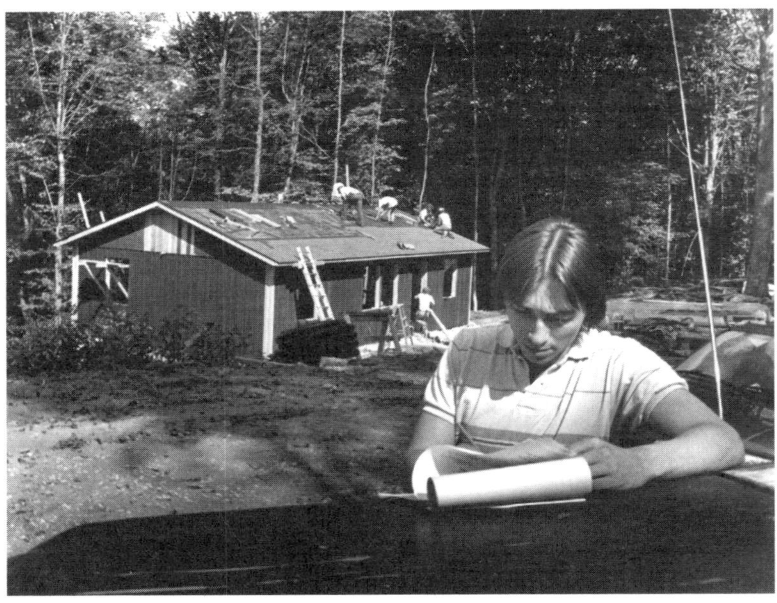

Fig. 21.1. A community center under construction near Clairfield, Tennessee

solar food dryers, and farm sleds. He directed the building and remodeling of eleven houses and numerous barns, outbuildings, fences, trellises, and pens. Wood, especially poplar, oak, and pine, was a creative medium, and in my family we learned almost by osmosis to use hammers, saws, wood drills, carving knives, and planes from an early age. We also became familiar with various types of wood and learned which ones could be easily fashioned, fastened, and polished. We also learned that it takes much work and a lifetime of practice to be a woodworking professional and that only the best craftspeople can enhance the beauty, native grain, texture, and color of wood.

Wood Treatment and Care

The great disadvantage of wood over other building materials is that exposed woods can decay easily and are susceptible to damage by termites, carpenter bees, and carpenter ants. To protect structural wood, a host of coverings, treatments, and techniques have been tried. Paint of various compositions is the traditional wood protector. Many chemical wood preservatives have been found to leave toxic chemicals in the nearby soil and potentially harm users. Most wood varieties do not last long when exposed to the elements, and thus protective eaves or good roofing and siding are structural defenses.

We have found that a combination of turpentine and raw linseed (or rapeseed) oil is a good nontoxic preservative. There are additional nontoxic methods to control or discourage wood-boring insects, including boric acid, which occurs naturally and is not volatile. Boric acid pesticides are available as baits, dusts, gels, and sprays, and as long as they are concentrated in areas away from human contact, they pose few health risks. Heat can also be used to eliminate pests. For residential applications, some thermal treatment systems use direct-fired and/or indirect-fired units capable of producing 4 million Btus per hour to eliminate destructive pests such as termites, beetles, and cockroaches.

In every remodeling project or new construction, low-VOC-emission adhesives for flooring, wall coverings, and woodworking that do not off-gas harmful emissions must be specified.

STONE IN APPALACHIA

Diverse types and sizes of stone are found in most of Appalachia. The ancient Appalachian Mountains are not as well known for their rock formations as are the jagged peaks of the Rocky Mountains. However, the region is famous for its cliffs, gorges, caves, and natural bridges. Most of Appalachia was not covered by glaciers, and thus the scattered, rounded surface boulders of northern, eastern, and midwestern parts of the United States are not common in that region and points south. Instead, much of the rock in central Appalachia is of sedimentary limestone and sandstone, which may be present in layers or broken-off chunks or carved from rock protrusions or overhangs. This limestone is beautiful but is vulnerable to chemical activity from acidic air pollutants, which can weather the stone over a period of time (though newly developed plastic coatings can retard or halt the stone degradation).

The early European settlers began immediately to use native rock in a variety of ways: as road markers, stone chimneys, foundations and walled buildings, roadside retaining walls, gravestones, tomb boxes to protect graves from foxes, and borders for walkways and gardens. By the late nineteenth century, rock fencing became common in parts of the region. The dry-stack rock walls that are typical of the region come in natural hues of grays, blues, tans, browns, reds, and yellows. (These heritage dry rock walls also harbor snakes, to the displeasure of some homeowners.) Well-fashioned stone buildings and bridge abutments are some of the notable features of early settlements in parts of Appalachia, showing the sophisticated work of early stonemasons and builders.

Stonehenge in England reminds us that massive boulders can be moved into precise places without sophisticated equipment, can be graceful, and can last for millennia. Modern stonework can be similarly enduring and graceful. Many worthwhile Works Progress Administration projects from the Depression era survive to serve as monuments to their builders, often simple working people who took pride in their accomplishments. The design of contemporary buildings, however, tends to limit the artistic freedom of stoneworkers, since stone, more than almost any other artistic medium, demands space for the crafter to sculpt a final product.

Construction-grade stones are still popular building materials. ASPI has many retaining walls, steps, rock seats, pillars for signs, walkways, garden and wetland borders, and structural foundations made from stone, mostly sandstone. Some of these were made from local abandoned sandstone chimneys, painstakingly carved back when people had more time for precise stonework. Modern stone or stone-faced buildings are very labor-intensive and can be quite expensive.

Many of us with a little patience and some masonry experience can lay stone, first in simple fashion and then in more structurally meaningful ways. We work according to our own creative expression. The skill and attention required for working in stone is a satisfying exercise. The worker should follow standard tips for proper lifting of stone, making use of legs rather than back and keeping the stone close to the body. Goggles and gloves are standard equipment.

The following are pointers given by a professional stonemason, Eddie Stallsworth, who also cautions that stonework is so involving that it's easy to forget about keeping hydrated and stopping to rest from time to time.

> *Discover and select the stone.* Prospecting for the right stones is an art requiring a critical eye. Know what to seek and where to look in old building sites, quarries, riverbeds, and abandoned buildings (nonheritage sites). Excessive removal from sites, even with the owner's approval, is never a good ecological practice.

> *Build according to availability.* Most people can't afford to build according to a set plan, so craft the product with what is readily available, modifying the final structure according to the materials at hand.

> *Design in advance.* The finished product should be in the mind's eye of the stone crafter. One must know the lay of the land before beginning construction. A stone wall may require a stepwise approach to utilize the earth in the best way possible. If creeks or trees are present, have them enter the final design with grace and beauty. Even if the design is exact before building commences, some freedom should be allowed.

Stack materials conveniently. Make sure your stones are nearby before beginning work. Imperfect stones can be chipped to smaller pieces or buried as filler rather than being given a prominent place in a wall.

Build with care and use proper tools. Stoneworking takes patience and time. That is why a novice builder could learn much from professional stone crafters. Acquire the right tools: chalk line with plumb bob and line level, mason's level, spade, stakes, pry bar, geologist's rock hammer, pounding or mason's hammer, chisels (point, plain, and toothed), and protective gear (steel-toed shoes, gloves, safety goggles and possibly knee pads). Wheelbarrows or garden carts are useful for hauling stones.

Start with easy, less visible portions. A good foundation must endure the test of time, and thus the periodic movements of the earth deserve respect. The foundation must be started below the frost line and needs to be deep and wide enough to support the heavy wall to follow. For rock bordering, shallower foundations will suffice.

Use good stones. Experts talk about "junk" stones that should be destined for filler. The novice will tend to use everything, and that is a mistake. The mind's eye helps fill the upcoming space involving critical material choices. Experts save choice rectangular stones for corners and ends of walls. They attempt to use better stones for difficult edges and corners and strive to interconnect the layers with stones as in regular bricklaying. The more overlapping and linked the stones, the stronger the corners.

Plan for drainage. Water must be diverted away from the foundation and submerged part of the wall. Water that collects behind a retaining wall can freeze and thaw, causing cracks and the breakdown of the finished wall. Install proper openings in the wall with drainpipe.

Cement properly. Apart from dry-stack stonework, one must test and use mortar of the proper richness. The final ingredients depend on type of rock, weather conditions, and other factors. Cleanup operations after construction include pointing (pressing mortar into joints and smoothing to release excess moisture), brushing, and washing the walls (muriatic acid is good for lime removal).

Maintain the structure. Rock walls often crack somewhere, no matter how experienced the builder. The earth shifts, limbs fall on walls, damaging floods happen. Regard rock-wall mending as a springtime chore, for nothing lasts forever in its initial state. In Cuzco, Peru, one marvels at five-century-old stone walls that have withstood periodic earthquakes, a tribute to the Incas' appropriate technology and expert stone fitting.

EARTH MATERIALS

Earth materials have been baked, rammed, pressed, or otherwise fashioned into shelter-related brick, blocks, and loose or granular mud forms since primitive times. Made inexpensively from local materials, many such structures are still being constructed and quite a few older ones are still occupied; almost all are quite comfortable and snug dwelling places. Earthen construction means that forests do not need to be cut for exterior walls, the construction materials are safe for people to handle and to use, the homes are affordable, many levels of skilled and unskilled workers can be included in the building process, and the finished product is relatively fireproof and destined to last, with reasonable protection, for centuries.

But the danger of radon gas, which decays into radioactive particles that can be trapped in the lungs, must be considered. In certain regions, including portions of Appalachia, an earthen home may accumulate these toxic materials, depending on the degree of ventilation and the manner of construction (see figure 21.2). A mere scan of a general Appalachian radon map will not tell the entire picture. An airy residence with circulating breezes may not have a problem, but a home built with earthen materials that lacks good ventilation could accumulate radon. The danger, especially to those spending much time indoors, increases where smoking is permitted inside the residence. While the actual degree of danger is currently unknown, scientists are more certain about radon risks than about risks from most other cancer-causing substances, according to the Environmental Protection Agency. Lowering radon levels in the home may cost from $800 to $2,500 if an experienced contractor is engaged, but this cost can be greatly reduced by installing a passive radon-venting system at the time of construction.

People choose to build underground structures for any of several reasons: such homes are snug and remain at approximately the same temperature year round, they are considered easier to guard from theft, they are not prone to tornado damage, and they require no external walls or supporting structures. The advantages of root cellars have already been discussed and apply here (see chapter 7). Owners of underground structures seem quite satisfied with their building choice and are very happy about their minimal heating and cooling expenses. But underground homes do not come without effort and cost. Building costs for a living roof (a reinforced roof containing earthen overburden) or exterior waterproofing may mount up, and proper moisture control is absolutely essential. An often unforeseen disadvantage is that residents may feel shut in and somewhat claustrophobic living underground.

Earth-bermed, or partly buried, buildings use the earth for insulation and protection from the winter winds. They are generally easier to construct than totally underground houses and

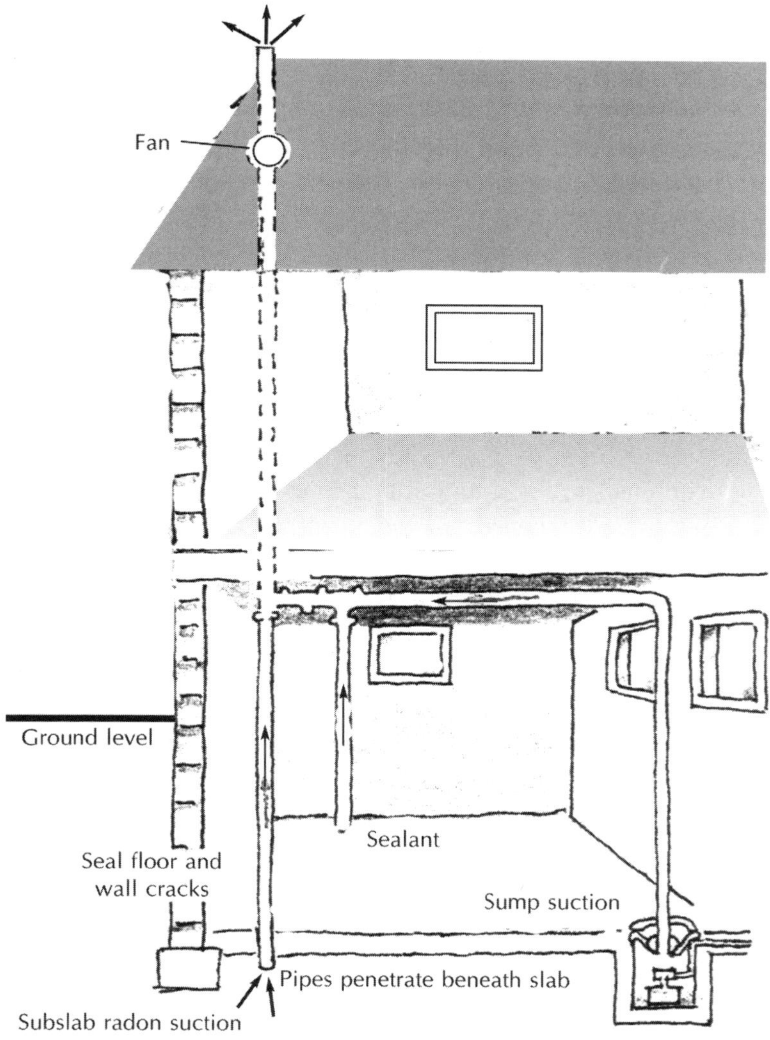

Note: This diagram is a composite view of several mitigation options. The typical mitigation system usually has only one pipe penetration through the basement floor; the pipe may also be installed on the outside of the house.

Fig. 21.2. Radon removal

can be designed to have adequate air circulation and provide satisfying vistas for mountain dwellers. The radon problem can also be solved through good design. These homes can be designed with piped, recirculating air that passes through the berm to serve as a geothermal system, assisting with cooling in summer and heating in winter. An upper story for viewing the landscape may prove more satisfying than a totally underground residence. One of the ASPI cordwood buildings is partly buried, adding to its fuel economy in heating and cooling.

Clay, which has been a local building material since soon after the initial European settlement of the Appalachian Mountains, is often baked into bricks of various colors and hues. Many people favor brick over wooden structures because of beauty, low maintenance, and long lasting properties. After the first pioneer homesteads were completed, wealthier people would make bricks at the homestead location from local deposits of clay when sufficient workers were available. Some of these homemade brick buildings have withstood the elements, while in other cases, where the bricks were not fired to high-enough temperatures or the clay was of imperfect composition, the buildings have been damaged or even crumbled in the ensuing years. While clay-baked brick was a material of choice for colonial American homes, garden paths, and street paving, the better grades of these products have always proved costly, for the kiln-baking process requires considerable fuel.

Today it is impractical to construct kilns and fire bricks on a construction site, and thus builders must resort to the purchase of commercially manufactured brick. Some home builders are located near brickworks, and for them the transportation costs prove far less than for those located a distance away. Radon problems are still present in brick structures and depend both on the building site and the degree of ventilation within the building. Brickwork of any sort takes skill and expertise, so it is generally not for do-it-yourselfers.

Rammed earth is a building material that is pressed into a mold of some shape. Rammed earth advocates speak of their structures as blending well into the surrounding landscape, being sound- and fireproof, having reduced risk from earthquake and wind damage, causing fewer health problems induced by poor indoor environments, and having lower lifetime operating costs for maintenance and energy use. David Easton of Rammed Earth Works in Napa, California, says that raw earth is commonly acknowledged as the world's most widely used building material. He speaks about the beauty of these buildings and how today earth building demands higher standards and greater care in construction than in the past.

Terry Green of Green Earth Builders in Carrboro, North Carolina, is interested in the Appalachian region. He affirms the same positive qualities of rammed earth structures, including beauty, energy efficiency, and resource conservation. Although these rammed earth buildings use a small amount of portland cement in comparison to concrete blocks, they are most often formed from large and heavy blocks of construction materials requiring backhoes, cranes, and other heavy equipment.

Tony Martin of Asheville, North Carolina, champions a new type of concrete called GigaCrete, which is only one quarter the weight of concrete and is extremely strong (8,000 psi). The material is fabricated from waste ash and a proprietary binding. The building components can be made to order and assembled with very little human labor or specialized skills. Housing costs are estimated to be far lower for this process than for conventional methods of making wooden structures.

A type of rammed earth structure called Earthship uses a self-sustaining building design that requires no outside source of water or electricity; cisterns fed by the rooftop catchment provide water, and photovoltaics provide cheap, abundant, and renewable energy. The building is built from castoff or recycled vehicle tires solidly packed with local soil. The north, east, and west walls are bermed with earth as an additional protective insulation layer. The south wall is made with insulated glass panels to use passive solar design. Stan and Pattie Jones have built what is perhaps the first Appalachian Earthship in Madison

County, North Carolina; they testify that indoor winter temperatures do not get below 60 degrees Fahrenheit, and summer temperatures are also moderate. Structure costs for the 2,000-square-foot building are running at roughly $40 per square foot; if labor were added, costs would be about $100 per square foot.

Building with cob, a mixture of clay, sand, and chopped straw, has been practiced all over the world for several thousand years. It has recently been rediscovered and used in various applications in Appalachia. The thick walls serve as excellent thermal mass and insulation to keep buildings naturally warm in the winter and cool in the summer. Cob buildings can easily be adapted to passive solar design to ensure ample warmth during the heating season. Various earth plasters can be used as finishes for the wall surfaces. Attention must be paid to designing sufficient eaves around the perimeter of the building to protect against driving rain and to ensure that the summer sun does not cause an overheating problem.

The building can be constructed in a barn-raising-style effort where everyone in the community, young and old, skilled and unskilled, can join to help sculpt the structure's walls. Raised doors, windows, and other building materials are easily woven into the cob structure's overall form. Potential cob builders do, however, have to take into account that mixing and applying cob can be a labor-intensive and time-consuming project. But all of the materials can come from local sources, and, after all, the building itself will be fully biodegradable, with no harmful residues left in the environment.

Because of the monolithic nature of cob wall construction and the walls' often curvilinear form, cob buildings are reputed to be highly earthquake resistant. To further increase earthquake resistance, John Fordice, an architect with the Cob Code Project in Berkeley, California, has suggested tying a concrete or wooden bond beam across the top of the walls to the foundation, with sleeved and tensioned vertical rods on 4-foot centers. Interestingly, a cob building is stronger than one made from brick or block because it contains no weak straight-line mortar joints.

PRESSED EARTH

Everywhere in my travels, from the coal fields of America to Botswana, to England and South America, I always found native building materials available. The resources were already there. The challenge was to bring available techniques, like the simple Cinva Ram, and then the education of the local labor force in its use. Added to this was the challenge of convincing the local business community that there could be a profit for them in developing the use of native building materials. Structures (albeit small, and small is beautiful) can be produced in large numbers and at low cost, if we set our minds and bodies to work on what we already have available.

Chris Ahrens, professor emeritus,
appropriate technology and international relations,
Warren Wilson College, Swannanoa, North Carolina

Building Materials

Pressed earth buildings have all the advantages of earthen structures, but they can be built without heavy excavating or hoisting equipment. People in developing countries with more available labor and little equipment can make earthen bricks with ease. Long Branch has conducted several hands-on workshops in the making of Cinva-Ram bricks with its hand-operated press and has constructed from them a Finnish contraflow masonry heater that has proved to be a highly efficient and aesthetically pleasing wood-burning device. For those looking to lighten their footprint on the Earth, the material excavated from a building site can be made into bricks, and the structure itself can be constructed from excavated earthen materials. With some coating protection from the weather, the structure will most likely remain long after the builders' lifetimes. Earthen bricks require basic material that can be obtained locally; they have high thermal-mass value as well as strength and enduring properties, allowing the structure's walls to be both self-supporting and heat-storing at the same time.

Earthen bricks have a number of superior qualities that result from the use of stabilizers and the rigidity derived from pressure applied to the blocks. Here are some advantages of this building method:

Earthen brick buildings are warmer and snugger in winter and cooler in summer because of the earthen mass.

Material costs for earthen brick walls are very low. However, there are certain labor costs that must be acknowledged and expected.

Suitable soils for earthen brick are easily found and modified in most areas of the eastern United States.

Earthen bricks are fireproof, and the sheer mass of an earthen structure gives comfort to its inhabitants.

Earthen bricks are made of the Earth, and the spiritual value of touching the Earth and harkening to our source of life is actualized in preparing and using these blocks.

Earthen bricks make excellent thermal storage for passive solar designs.

Earthen-brick buildings are better built with many hands helping and are thus ideal subjects for communal projects, a major aspect of an appropriate technology application.

The environmental cost of earthen brick is very low, for few trees are needed. An African program builds earthen brick houses with fiber-concrete tile roofs, using only half a tree per house, which they compare to the twenty trees' worth of energy needed to build the same house from burnt bricks.

Cordwood takes even more forest material, though much of it can be from trees removed for other projects or from recently downed timber.

Earthen bricks are twice as strong as ordinary baked bricks.

Walls of earthen brick are very attractive. If desired, colorants can be added to the mixture to produce various hues.

But there are some disadvantages as well:

Processing and completing earthen bricks takes time and is labor intensive. Comparing the earthen-brick-type building with cordwood (chapter 22) reveals that cordwood construction takes less time and care in preparing the basic cordwood and has many of the advantages of earthen brick construction. However, not all places have such a generous supply of wood.

In earthquake zones, earthen brick houses are not safe unless used to infill for timber frame or poured-concrete post and beam construction. This same disadvantage applies to cordwood and other such buildings. While many regions of America seldom experience devastating earthquakes, builders may not want to bear the risks in areas where earthquakes are at least remote possibilities.

Curing time for the earthen bricks must be allotted, and thus planning should involve working on foundation or grounds while the wall blocks are drying. or staggering batches to allow drying time while construction of parts of the walls and roof continues.

The exterior wall surface must be protected from the direct impact of rain or splashing from the roof. Extended eaves generally will provide this protection.

Radon can be a problem, although this can be addressed with proper ventilation techniques.

Building Techniques

The making of earthen bricks is an ideal Earth-friendly technique because of low material cost, low demand for specific building expertise, the accrual of value through sweat equity, the use of local materials, and the fact that the project lends itself to being part of a community project by people with limited resources. This section offers some basic information worth remembering in construction.

Foundation. The foundation is built in standard ways using concrete or stone with mortar and extending a foot or so above grade. The width should be the same as the block to be used in the walls. A poured concrete floor is more expensive but takes far less time. However, some builders prefer to use the same machine that makes the block to turn out thin floor tile much in the same fashion as the earthen brick walls, and this can be laid after the main structure is complete. If the foundation is not flush with the walls, it could catch water and erode the block over time.

As with any building, all surface water needs to be directed away from the foundation. Thus suitable attention should be given to extending the eaves, installing proper guttering, and providing for ground drainage.

Soil for earthen bricks. This could be excavated from the foundation portion or the subsurface of grounds on which the building will stand. High-quality topsoil and organic matter should be removed and the topsoil transferred to a future garden spot. The subsoil should be 10 to 25 percent clay, sifted through a ¼-inch wire mesh affixed to a sturdy 4-by-4-foot frame suspended hammock fashion from two trees, or a portable frame. Sifted soil must be kept dry.

Soil testing. Subsoils are mixtures of sand, clay, and silt. Knowing the proportions is important for an optimal earth product. The ratio of the three basic components has some flexibility, but proportions should be about 75 percent sand, 15 percent clay, and 10 percent silt. If the mix is less than 10 percent silt, more does not have to be added, but it may be important to add sand in high-clay-content soils (with the clay component being no more than 25 percent). In parts of the eastern United States sand must often be added to bring the subsoil to the right composition. Actual soil composition is determined by taking a wide-mouth glass jar and filling it halfway with sifted soil. Put on the lid, shake vigorously for three minutes, and allow to settle overnight. The soil will settle in three ingredient layers with sand on the bottom. In rare cases, the clay will be deficient and can be augmented by mixing in Bentonite or other clay soils in the vicinity.

Mixing and stabilizer. A measurement of shrinkage is also needed to determine the amount of stabilizer to add to the soil mix. Construct a 2-foot (interior dimension) wooden trough that is 1½ inches wide and deep. Moisten some of the soil mix so that it just barely sticks together (but not to one's hand) and pack it firmly in the trough. Allow this to dry in the sun for three days. Push the dried contents to one end and measure the gap or shrinkage. For ½ inch of shrinkage use 1 part stabilizer to 18 parts of soil, for ½ to 1 inch use 1:16, and for 1 to 1½ inches shrinkage use a 1:14 ratio.

The materials may be mixed in a standard concrete mixer or manually in a wheelbarrow using a shovel and concrete hoe. Follow routine mixing procedures, first thoroughly mixing the dry materials, then adding water and mixing. When properly mixed, the materials will have the same appearance and consistency as in the shrinkage test. The proper consistency can easily be gauged by ordinary masonry workers.

About 5 to 7 percent of the earthen bricks will be stabilizer. One type of stabilizer is portland cement, which is expensive, takes energy to produce commercially, and ordinarily is transported from outside the region. Lime, also an efficient stabilizer, takes less energy to process. Use slaked or agricultural lime, portland cement's main ingredient. The lime takes twice as long to cure—two weeks wet cure and two weeks dry—but makes a stronger block. In four more weeks the block will reach maximum strength. The lime-soil mixture should be allowed to stand for one or two days before block-making operations. Portland cement when used as a stabilizer is mixed moist and used immediately. This cement mixture will wet cure in one week and dry cure at maximum strength after two more weeks. Some prefer a cement/lime combination that is made by first mixing the lime as above and covering for one to two days and then adding cement and moistening to proper consistency.

Block making. The most important tool for block making is the block press. One block press is the Cinva-Ram, which is imported from Bogota, Colombia. A similar power-operated device is now available from an American manufacturer; more details are available from Habitat for Humanity. The press is filled with proper amounts of soil mixture (130 pounds pressure on the end of level). If water drips from the block, it is too wet. The inside of the press must be oiled before use to allow the block to come out without sticking.

Curing. The extracted blocks should be stacked on a clean, flat, level, shady surface. One possibility is to build a post and beam structure with the roof already constructed so that the floor of the building serves as a shady curing area; here the heavy block will not be carried too far for construction. If the walls are to be supporting, a roof could be constructed with removable supporting posts just off the foundation to allow the wall-building process to continue after block curing. In the drying process, these pressed blocks should be stacked not more than five high with air space between all blocks and each layer in alternating directions. The blocks should be covered with a cloth during the wet-curing period and sprinkled with water each day in nonhumid climates. The stacks should be dry cured uncovered in the open air for the time specified by the stabilizer manufacturer.

Wall construction. The building of the walls requires the same expertise as laying concrete block and brick. It is best to have an experienced bricklayer at least initiate the operation and offer pointers if all the builders are novices. The earthen bricks are essentially 6 by 12 inches, 4 inches thick, and require ½ inch of mortar. The mortar can be made with various amounts of cement and/or hydraulic lime. An expensive prepackaged mortar such as Internee brand type-N Coplary cement can be used in the same manner as for preparation for brick- or concrete-block-laying operations.

If post and beam (wood or concrete) construction is used with the roof already finished, a single thickness of block should be used for the wall. A layer of rigid foam should be placed on the exterior of this single wall, followed by a stucco coat. If the pressed earth block is the supporting wall, it should be laid in a double layer and connected with brackets, with rigid foam between the layers. Another alternative is to dry-stack the pressed earth blocks with a ⅛-inch layer of fiberglass bonding cement plastered on both interior and exterior walls.

Wall building around windows and doors should follow standard bricklaying techniques. Moisten all blocks before applying mortar. Build the corners first using string and a level. The block can be easily cut with a brick mason's hammer. Window sills could be of poured concrete, fashioned so that water is shunted away from the walls. Wooden lintels should be set over the windows and door apertures and extended past the sides for one foot on each side. A binder beam of reinforced concrete could be poured around the top perimeter of the wall.

Completion of construction. Once the binder beam is cured, a roof may be added if it was not built prior to the completion of the walls. Light-colored or reflective galvanized or aluminum metal roofing is best for minimizing heat gain during summer months. Cedar shakes, often imported from the West, are attractive and natural, but they are combustible. It is possible to make roofs from tile, but the tile must be glazed and is quite heavy, demanding more rafter and roofing support. As in cordwood buildings, the wall could also be the interior surface. As mentioned, this could be tinted to give color to the house. Some prefer an interior finish like a brick house with furring strips and sheetrock or a plaster finish. The exterior wall could be covered with stucco for weather protection.

A structure made of earthen brick is built to last and, when properly constructed, should be usable for a thousand years or more. As with a cordwood building, the thicker the walls, the more roof and floor area is needed to achieve a given amount of living space. Narrower walls mean more living space, so the post and beam single layer of earthen brick could be preferable.

When researching building costs in 1996, we found that an earthen brick cabin of 400 square feet would cost about $2,000 for materials for the basic structure—about the same as a comparable cordwood building. Cement costs have increased since then, of course; in 2006, the estimated cost is $3,000. The toilet and utilities are extra. The assumption in both cases was that doors and windows were salvaged and that the interior would have earthen rather than plasterboard walls. The floor was assumed to be pressed tile in the case of the earthen brick but poured concrete for the cordwood building.

While the materials cost per square foot for the two types of structures are similar, there is a great difference in the time required for construction. I was able without assistance to build the walls of the 1,000-square-foot cordwood structure in two weeks, after the cordwood was gathered and cut by several persons over a three-day period. The testing, sifting, mixing, curing, and mortaring of the pressed earth materials takes about two months. Remember, though, that the cordwood structure takes ten or more cords of wood, whereas the earthen brick structure takes virtually no wood apart from the roofing system.

CHAPTER 22

Cordwood Structures

After living in the Cordwood House through a little more than one full seasonal cycle, I have come to appreciate the ambiance the stacked walls offer. The house is quiet, and easily cooled in summer. In winter, although I heat with wood and do not maintain a fire during the day while I'm gone at work, it is still warm when I return.

Deb Bledsoe, director, ASPI

Rural America, especially in the forested eastern portion, has used its native wood for a wide variety of building purposes: red-oak-framed tobacco barns, chestnut-sided homes, and cedar shake roofs. While some of these have passed, or are passing, from the scene, native wood is still available and is a long-lasting building material when protected; it continues to be in favor in home construction.

Widely used in the abundant forests of Scandinavia, the log cabin came early to Colonial America and has remained a relatively popular style, with many such structures continuing to dot the Appalachian and eastern American landscape. A modification of the stacking of horizontal logs for walls is the stackwood or cordwood building, a Northern European building technique. Debarked logs are cut into the desired stacking length (e.g., 8, 12, or 16 inches) and cemented in place with insulation as a thermal break between interior and exterior mortar joints. The cut-off ends of the logs form the interior and exterior facing of the structure. When carefully placed, the wider stacked walls can even be load-bearing.

The origin of the cordwood building is lost in the hazy Eurasian past. Greek structures are known to be over a thousand years old. Northern Europe has numerous examples of heritage structures, and some are found throughout Siberia and across into Canada. They are known to be quite snug during the bitter winters in those frigid lands. Cordwood buildings require little fuel wood to heat in winter even when outdoor temperatures dip far below zero, and they are amazingly cool in the summer. While some buildings are built mainly for winter or summer use (see chapter 20), the cordwood building is truly an all-season structure.

Why, then, aren't cordwood buildings even more popular in forested areas that are subject to temperature extremes? One factor is that conventionally framed houses, which are relatively wood conserving, became increasingly fashionable in the nineteenth century, beginning in areas where wood had become somewhat scarce and spreading elsewhere. Furthermore, cordwood buildings have been made from cedar and other evergreen stock, thus further limiting building material. Another problem is that more modern houses have been built to follow fashion and, unlike in older times, without much regard to what building materials are locally available (see chapter 21). Both home buyers and housing lenders prefer to build in a uniform manner from Alaska to Florida with some insulation modifications.

ADVANTAGES OF CORDWOOD BUILDINGS

Cordwood building structures have a number of advantages:

Fuel savings in both summer and winter. The cost of the extra wood required for construction of cordwood houses is recovered in about three years through radically reduced demand for fuel wood (one cord per year for a 1,000-square-foot cordwood building versus three or four in conventional ones).

Very low construction costs. On a per-square-foot basis this is one of the most economical houses to build, especially in areas where there is plenty of wood.

Use of recycled materials. The short size of the cordwood log (8 to 16 inches) allows scrap and log ends to be saved and used. One ASPI cordwood building is made from pine waste donated by a post-making establishment.

Mold-free walls. Cordwood buildings do not harbor mold and fungus as do some that use gypsum board and straw bales.

Beauty. Cordwood buildings are aesthetically pleasing structures.

Easy construction. Full-length logs are heavy and cumbersome, difficult to transport with draft animals or heavy vehicles, and require strong people, if not cranes, for lifting and placing. A cordwood building can be built by a single person with moderate stamina and strength. Care is the only skill required.

Easy maintenance. A cordwood building requires only a modest amount of maintenance compared with a building that needs frequent repainting inside and out.

Low fire hazard. It is almost impossible to burn down the walls of a cordwood building, even though the roof and interior furnishings may be lost. The walls are protected by the surrounding cement from all but singeing in case of fire. This makes it an excellent structure for forested areas, which are fire-prone.

Low-tech tools. Modern frame buildings require many power tools, but few are needed for cordwood buildings except foundation tools, a chain saw to cut wood into lengths, a hatchet for debarking (if needed), wedges, a splitting maul or "wood grenade," masonry mixing tools, and other tools for the interior or roofing.

CONSTRUCTION

The classic books by Rob Roy (see Resources) describe cordwood building construction in great detail. Builders are often advised to build a small companion cordwood outbuilding to acquire skills and overcome mistakes before going to the main structure.

Homestead site selection is stressed in other sections of this book, for good reason. So very much depends on that original siting decision for the total comfort, ease in building, future access, and lifetime economy of the building. An accessible building can be appreciated by more people, thus popularizing new and innovative appropriate technologies. Southern exposure is quite important, as is proximity to water, shading, gardening areas, and the presence of windbreaks to protect the house from harsh winter winds.

Most easterners are limited by the availability of wood in quantities needed (six to ten cords) for a moderate-size cordwood building. Traditionally, cordwood buildings have been constructed in northern climates, and the eastern red cedar *(Juniperus virginiana)* has been the wood of choice. Cedar is ideal because it is generally plentiful in logs of small girth, durable, and rot resistant. ASPI used pine scrap for two structures and white oak *(Quercus alba)*, a common tree in central Appalachia, for its first structure. The "recycled" downed oak was taken from tornado-damaged, cut-over U.S. Forest Service land only a mile away. Another relatively plentiful tree that resists rot when protected is the tulip or tulip poplar tree *(Liriodendron tulipifera)*. Logs of this abundant eastern American tree have been known to hold up well in two-hundred-year-old log cabins. Thus the required wood may be either evergreens or construction-grade hardwoods protected by overhanging roofs. The porch, a traditional feature of

Appalachian cordwood buildings, helps protect the portion of the building exposed to the harshest weather.

A post and beam structure is often used for cordwood buildings, but this takes extra framing time. Such a step is omitted when the cordwood wall becomes the supporting structure. However, the walls of the cordwood building must be thick enough to provide good support. A rough calculation will demonstrate to planners how exterior shape has much to do with the amount of wall per interior space. The same exercise could be performed to demonstrate wall thickness and the size of the final living space. The amount of space can be maximized in relation to the amount of wall through greater circularity and the cutting of corners. Thus a round building has the least length of wall per space gained, followed by an elliptical building, a square design, and finally a rectangular building. Moreover, every inch counts when calculating wall thickness. Thicker walls are better insulated but require more roofing and flooring. The oblong ASPI building was built rapidly up to completion of the walls. The cutting and fitting on the roof made me declare that I would never again build a structure of this shape.

Debarking is necessary to prepare the logs, and the degree of difficulty of this operation may help determine the type of wood used. If the logs are stacked in ricks for a minimum of a year, the bark of many tree types will loosen and can be removed quite easily. Some varieties can be stripped of bark immediately after cutting. Cutting into final lengths may depend on the desired thickness of the walls. A load-bearing 16-inch-thick wall has greater insulation value (R is generally 1 per 1 inch), but it takes up more roof and floor area, leaving less living space. If the logs are fillers in a post and beam structure, less cordwood log thickness is needed and the log length can be as short as 8 inches. However, overly thin walls are harder to lay and hold in place and may require ties and binding. Splitting of logs that are 10 inches or greater in diameter will reduce checking or cracking of the logs once in place, although checked logs can always be caulked. Finer-split logs take up more space, meaning less wood is needed, but such logs require more cementing.

A masonry or slab concrete floor will retard termite and other insect damage. Standard protective measures apply here. One advantage is that the wood wall is not continuous but is separated by cement and insulation around the stacked logs. Stacking of the cordwood should begin a foot or so above the grade on a masonry foundation. Aluminum flashing may be applied between the foundation and the cordwood to discourage crawling insects. For additional insect protection, a mixture of Thompson's Water Seal or other wood-preserving materials should be used, especially on the lower layers. A solution of half turpentine and half linseed oil should be applied to the wood during construction, then reapplied during maintenance operations in future years. Aging oak walls invite wood bees or small burrowing insects to lay eggs in some of the logs. The preservative application discourages insects for a year, but they tend to return, although they don't cause major structural damage.

As for cementing, Roy suggests using mixtures of materials plus ample amounts of sawdust as a cement filler. However, since shrinkage rates depend on the type of wood used and the curing time, no formula is right for all situations. If the wood is thoroughly seasoned, shrinkage is less of a problem. We have found an ideal outer-coat mortar mix to be fine white sand and Brixment brand type-N Coplary from Essroc Materials Speed, Indiana. This mortar mix clings well to the wood surface provided the wall is not subject to hammering and other violent vibrations. Cementing inner logs with some sawdust mix could be followed by tuck-pointing using the sand and Brixment mixture to maximize insulation and minimize cement costs. The inner half to two-thirds of the space between stacked logs should be filled with scrap fiberglass insulation, and thus the mortar should be applied only to the outer and inner 3 inches. In other words, the mortar is put only on the outer and inner sides, not within the middle portion of the stacked logs. (See figure 22.1.) The more insulation between the mortar joints, the

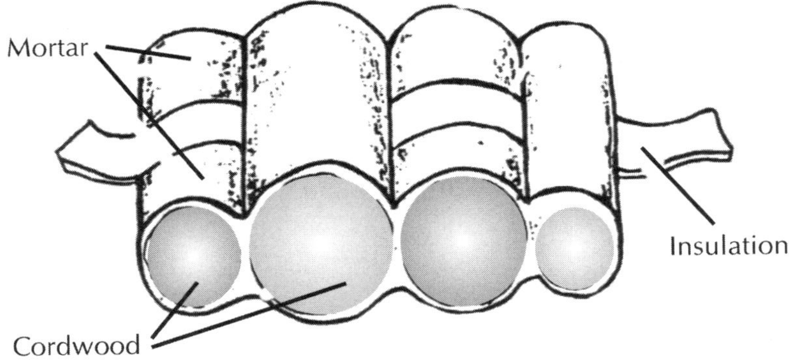

Fig. 22.1. Insulating a cordwood structure

greater the overall insulating efficiency of the building. Scrap insulation should also be used between the top plate holding the roof and the roof itself and between window and door framing and walls.

Much of the finishing work of a cordwood building follows standard plumbing, electric wiring, and interior-wall practice. The roof of curved buildings takes more time to build than the walls. Had the roof of our oblong structure been extended in a Swiss chalet or gable fashion over the oblong walls, the roof problems would have been greatly simplified, resulting in a need for less roofing material, a shorter construction period, and greater roof protection of the end walls. A roof overhang of 2½ feet will generally protect the walls, depending on the height of the building and the direction of prevailing winds and rain. Another protective measure is to construct an adjacent outbuilding in the direct path of the prevailing winds.

ASPI's addition of cordwood walls to an aging mobile home (figure 22.2) has helped make the structure far more attractive and increased its resale value. Manufactured homes make up nearly half of modern rural housing in America. These structures generally depreciate in value over time, while the value of most permanent houses remains stable or appreciates over the

Fig. 22.2. ASPI cordwood/mobile home

same period. Cordwood walls greatly improve insulation of the structures and give them greater permanence.

MAINTENANCE

Keeping cordwood buildings in good shape is not an onerous task.

Here are some maintenance tips:

Remove wasp nests from the prominently overhanging eaves. While wasps are good in the overall biological scene, their stings can be painful and some people are allergic to them. Kerosene is an excellent low-toxic removal agent for it knocks out wasps quite easily.

When attaching pictures or other objects to the cordwood building interior walls, be extremely careful in using a hammer. The masonry joints on the walls can be damaged by heavy blows.

As already mentioned, it is a good idea to apply a solution of half linseed oil and half turpentine every few years to the exterior and, more rarely, to the interior.

Make an annual inspection for cracks and use caulking or cement for the repairs.

SPREADING THE WORD

Becoming convinced that the cordwood building is an excellent form of rural appropriate technology and taking the steps to build such a structure are only two parts of the total picture. Hidden examples of innovative housing are not sufficient for bringing such structures to the attention of others. A better procedure is to maintain and demonstrate the structure. As with solar applications, communities need to promote this low-cost Appalachian housing and use such buildings as demonstration projects. This has been done in Rockcastle County, Kentucky, with three such structures only two miles from exit 49 on Interstate 75, Appalachia's busiest thoroughfare.

A number of individuals and groups have hosted workshops when cordwood buildings are being constructed. This gives inquisitive builders the confidence to build even when such buildings are not yet in fashion. A Continental Cordwood Conference was held in 2000 in New York.

Other ways of spreading the word include a growing literature on cordwood techniques, including books, technical papers, and videotapes. But the best advertisement is a satisfied resident who is willing to show her or his house.

Telling the construction story is not complete without critical comments about the structure presented. If we were to rebuild ASPI's oblong building mentioned here, it would have some colored glass bottles inserted for beauty and stained glass lighting effects; it would have some earthquake-proofing supports; it would have more windows and perhaps skylights; it would have a gabled roof; and there would have been more splitting of the logs. The best story includes both the advantages and the imperfections, and cordwood building advocates find little to hide and much to celebrate in the long run.

CHAPTER 23

Yurts in Appalachia

> *You have noticed that everything an Indian does is in a circle, and this is because the Power of the World always works in circles, and everything tries to be round. . . . The Sun comes forth and goes down again in a circle. The moon does the same, and both are round. . . . Even the seasons form a great circle in their changing, and always come back again to where they were. The life of a man is a circle from childhood to childhood and so it is in everything where power moves. Our tipis were round like the nests of birds and these were always set in a circle, the nation's hoop, a nest of many nests where the Great Spirit meant for us to hatch our children.*
>
> Black Elk, Ogalala Lakota (1863–1950),
> *Black Elk Speaks*

The yurt, a circular domestic dwelling (house or tent), is traditionally found on the rural steppes of Mongolia. In that part of the world the structures are made from animal skins and poles and held together by cords wrapped around the circumference of the building just below the eaves. The cords hold the walls and roofing rafters firmly in position, much like barrel hoops hold staves in place. Thus the yurt is as rigid as any traditionally framed building but made in a far simpler manner. For Mongolians, this structure is either permanent or temporary and was the only known type of building until recent times. It continues to have deep significance in that culture and "provides the space in which every category of person or object in the nomad's world could be located, and so becomes a kind of microcosm of the social world of the Mongols."[1]

Is a type of building developed by a nomadic people thousands of miles from Appalachia an appropriate housing model for this area? Concepts of affordable housing are always of interest, even if different materials must be substituted. Construction materials are costly when they must be imported from distant places, but ideas travel easily and, with some modifications, can be quite relevant for another cultural area. The yurt is a testimony to the fact that appropriate technologies do not just spring up in the developed world. The yurt is a great design gift from another part of the world and is offered to those who may think they have or know everything. Appropriate technology is not something imposed by a colonialist mentality but rather an egalitarian sharing of ingenious, resourceful, and practical ideas that can be communicated at little cost and applied across the globe. A faddist approach to structures. in contrast, may lead to application in unsuitable places of a technology that was a good idea at its place of origin (e.g., construction in humid climates of straw-bale structures that are quite suitable in drier places).

American building practices borrow more heavily from other lands than is generally admitted. Construction methods for log cabins come from Sweden, and many of the permanent structures in towns and countryside incorporate European building concepts. Why not borrow from Asia? Many Americans are modern nomads in their migratory lifestyles and can explore various temporary or semipermanent shelters. Also we are often seeking beauty, harmony, and centeredness in a world full of centrifugal forces, and an aesthetically pleasing circular building can fit easily into rolling hills or mountainous landscapes.

A dozen yurts, one of which is quite large, have been built by the Woodlands Mountain Institute, an educational institution in eastern West Virginia's Pendleton County. The institute is in an

1. Caroline Humphrey, "Mongolian Yurts," *New Society,* October 31, 1974.

incredibly beautiful upland setting within the Monongahela National Forest and is an ideal place for nature courses for elementary and high school students, with college students serving as interns. The goal is to get these young people to appreciate the biodiversity of the region and at the same time to have a relationship with peoples from other lands and cultures. Thus Woodlands has an international flavor with overseas programs in the Himalayas and Andes. The yurts help prepare students for their overseas experiences. These, along with other, more traditional, structures, serve as lodging and training facilities, all with conservation of resources and ecological principles in mind.

The community center yurt has served for a quarter of a century as the nerve center of the institute's operations. It is an extraordinary two-story structure with concentric round rooms in the center that fit seamlessly into the mountain environment. The first floor is used for domestic operations; the room above serves as a dorm space. Good natural lighting is provided by the windows at the peak of the yurt. The roof is made of local cedar shakes and the native wood siding is from local lumber mills. The interior has a homey hardwood finish. The perimeter of the major interior room is a study and work space, with room also for storage of materials and for book displays. Many of the other yurts are for student and intern sleeping quarters and storage space.

ADVANTAGES AND DISADVANTAGES

Yurts hold immense promise, especially as auxiliary buildings. Some advantages include:

Low cost. A yurt can be built as a permanent structure using inexpensive, locally available materials for only about one-third of the cost of a conventional wood-framed building. A 133-square-foot building can be built for about $6 per square foot using purchased new materials; if used, salvaged, or recycled building materials are utilized, the cost could be far less. However labor costs, if donated labor it not used, could mount, depending on the degree to which the builder desires to fine finish the structure and what kinds of interior finish materials are selected.

Speed of construction. Yurts were taken down and transported across the steppes frequently and rapidly. The yurt can be built in a short time because the frame does not have to be permanently fitted in place. A more permanent building can easily be built by two people in a week, though more help may be needed for the major raising when the circumferential steel holding cable is tightened.

Durability. The rugged and flexible structure can withstand major shocks from earthquakes better than most masonry buildings. The basic structure of the ASPI yurt (figure 23.1) withstood a major windstorm in 1992, but there was some damage to the roof, in part because the holding ring at the crown of the roof was not properly fastened. With that exception, the ASPI yurt weathered the event better than a nearby frame building, which the storm moved off its foundations.

Ease of maintenance. Its very simple design means the yurt requires little maintenance. All areas of the building are accessible, with no hidden corners to collect dust. Because of the low profile of the building (most yurts are just tall enough for people inside to stand upright), extreme dangers for workers climbing on roofs are eliminated. Overhanging eaves are included in the design to protect the building's siding from weather. The only difficulty is that wasps like to congregate beneath these eaves and need to be removed so as not to threaten residents and visitors.

Fig. 23.1. ASPI yurt

Snugness. Because of their round shape, yurts instill a feeling of being truly at home—warm and snug at the center of the universe. They have smooth and rounded sides and no complicated corners like those found in geodesic domes, which often prove difficult to weatherproof. Yurts are ideal dwellings over the short term for those wanting a hermitage experience or for hikers, visitors, or vacationers in Appalachia.

Roomy feeling. When one enters a yurt, there is a feeling that the space is far larger than it actually is. This sense of spaciousness is due to the expansive shape of the building and to the fact that much of the natural lighting is from a skylight.

Attractiveness. The architecture of the building is really pleasing to the eye because it blends in so well with the natural curvature of the landscape.

Why hasn't the yurt caught on? Building these structures requires people with some construction skills, and this limits the number of willing do-it-yourselfers. It is best to assist another in building his or her yurt first before undertaking the construction on one's own. Here are some other possible drawbacks:

Foreignness. To some a yurt might resemble a temple or the office of a foreign agent. That kind of observation is not heard as often now as in past years, but it may be enough to dissuade some perhaps less open-minded builders from going ahead with a yurt. Will others be willing to explore the possibilities of building these structures down the road?

Incompatible shape. One concern is that the yurt cannot easily accommodate traditional rectangular furniture, but this objection may be principally in the mind of the beholder. Permanent shelves and sitting places can be easily installed along the structure's inner walls. Cushions, futons, and bed mattresses work very well as alternatives to rigid sofas and beds.

Negative perceptions. Some remark that neighbors and friends may not immediately understand or appreciate the buildings. Yurts aren't the only types of buildings to face ill-informed criticism; peer pressure has led to the proliferation of unfriendly and climate-insensitive housing designs from Alaska to Florida, thus severely limiting the capability of our built environment to reflect the distinct heating and cooling requirements of specific bioregions. Unimaginative designs contribute to the monotony of the landscape and diminish the creativity of these buildings' inhabitants.

Window placement. Some complain that there is not enough lower window space on the exterior wall. Skylights and over-head windows give a sense of light and freedom to enjoy the vast expanse of the sky dome, but those who need additional light can have doors with window panes in them or additional windows, which should be installed with proper care being given to preserving the curvilinear shape and structural integrity of the exterior walls. In fact, quite often people prefer the privacy that fewer windows provide. The better approach to achieve more contact with nature is to build decks, platforms, porches, or outdoor pavilions.

Construction work time. A final concern is that the amount of work required per square foot of living space for a permanent yurt (not a tent variety) is a little more than for the average wooden frame construction, even though yurt construction does not demand framing expertise. It is true that the greater savings are realized in the limited number of building materials that are required. As one might expect, yurts present excellent opportunities for those conducting building design workshops, communal or barn-raising building activities, or projects by groups of do-it-yourselfers.

VARIETY AND UTILITY OF YURTS

The yurt, in Mongolia and elsewhere, is generally a single-room structure, but multiroomed varieties are possible. In an inner circle within the yurt, occupants gain a feeling of well-being and achieve a sense of unity. Yurts have creative space where all tasks, including cooking, eating, sleeping, relaxing, and working in harsh weather can be done by the original designers and builders. Yurts are ideal for retreats and for periods of reflection.

As mentioned above, the creative space may be enlarged when more people share the yurt. By enlarging the yurt with additional rooms and turrets, designers can obtain more private space, which is so important to modern Americans. Design modifications can be as elaborate as the builder desires. All walls, floors, and ceilings can be insulated and made energy efficient for winter. Electrical wiring can be installed with no difficulty, solar applications may be added, and the building can become a comfortable year-round living or work space.

While Mongolians construct yurts using easily transportable hides, poles, and ropes, Western builders make them from materials at hand, following the principle of using native materials whenever possible. The ASPI yurt siding is rough-cut pine lumber, and the steel binding cable is recycled material. The roof is the key to the yurt's lasting beauty. At first, we used commercial plywood for the roof with accordion-type rib/rafters as designed by William S. Coperthwaite's Yurt Foundation of Bucks Harbor, Maine. We found that the carefully painted plywood was unable to withstand the high humidity of the Rockcastle River valley and the laminated layers became unglued. (More recent plywood adhesives are not water soluble.) A second similarly designed yurt roof was damaged in a violent windstorm, so we redesigned the roof using scrap pine slats from the local Kentucky Forest Products fence post factory. These slats were covered with heavy tar paper roofing, The roof has proved satisfactory for fourteen years, though it is not as aesthetically pleasing as the Yurt Foundation's style of roof.

CONSTRUCTION TIPS

Specific directions for construction are available from the Yurt Foundation (www.yurtinfo.org/yurtfoundation.php). Some points worth noting are summarized here.

> Siting is the most important decision. People try to obtain a scenic site that is accessible and has some protection from the winter wind and shade from the sun to keep the place cool in summer.
>
> Yurt foundations can be either poured concrete (if the necessary material can be easily brought to the site) or treated

wood posts placed in a manner that allows for a crawl space underneath. The yurt itself may rest on a platform made in the style of standard porch decking.

Flooring could involve installing insulating material for use in winter. The ASPI yurt has a plywood floor that is completely covered by a rug to hold in the heat.

Walls can be made from any suitable building or salvaged material at hand (local wood, shingles, etc.). ASPI used 1 by 12 inch rough-cut pine boards attached as an inner and outer wall in a stepped fashion. Scrap fiberglass insulation was inserted between the inner and outer walls.

Cables should be made of a strong, durable binding material that will last. Generally, for the sake of long-term durability, the cable should be galvanized steel at least ⅜ inch thick. In turn, these cables should be held in place by fastened rings or heavy-duty staples and tightened by cable clamps.

Roofs with insulation between inner and outer walls can be constructed for a winterized structure. At the top of the Yurt Foundation–designed roof, twenty-three ribs are held in place by firmly fixed compression rings. Such a roof is somewhat time consuming to install. We found our pine slab roof was less expensive but resistant to wind damage and simple to install, with three people completing the roof in less than a day. Its appearance is pleasing from the interior but less so from the exterior.

Windows can be added as desired, but they create complications. Yurts designed with ribbed plywood can have triangular windows at the outer end of the ribs, which give a ceiling-lighting effect to the entire building. A skylight at the center of the roof's cone is a standard way to provide light;

it replaces the smoke hole in primitive dwellings. The ASPI final skylight was a four-sided cap design that is framed with 2 x 4 pressure-treated material to withstand the wind. Cutting apertures for windows in the exterior walls may weaken the structure of the building, which is held in place by planking (like barrel staves). Additional window space can be added to the door itself or the entire door can be made from translucent material. Some builders prefer to use Plexiglas in place of glass, but this petrochemical product is quite expensive.

Insulation can be attached to the floor, walls, and roof of buildings meant to be winterized. Double-glazed and fitted windows and doors may be used to conserve heat. It is estimated that winterized buildings will cost about 40 percent more than those intended for summer use.

The yurt's interior may be finely finished and decorated to taste with trim and borders, built-in shelves and cabinets, and installed benches for seating that can also double as storage containers.

Yurts have been used or proposed for use in a multitude of ways in Appalachia: for permanent dwellings, libraries, storage spaces, adjunct bedrooms to houses, nonprofit office spaces, small variety shops at public or private parks, hermitages for single people or couples, shelters in remote areas or along nature trails, meditation rooms, overlooks at prominent observation points, bunkhouses for overflow customers, nature sleeping experiences for families, tool rooms (concrete floor desirable), small chapels, or learning centers at educational institutions. They serve a useful auxiliary purpose within a homestead or community and will fit those uses quite well. Determine the reason for building the yurt, and then build it to design and size.

CHAPTER 24

Simple Modes of Transportation

In 1997 Matt Green and I stripped a Dodge Colt of its dirty and complex gasoline system, comprised of engine, radiator, carburetor, and exhaust components. In its place, we installed a surprisingly simple, quiet and clean electric propulsion system comprised of an electric motor, motor adaptor, motor controller, and recyclable batteries. Later, and with help, I installed a solar charging station on our office building for the vehicle, which today on average delivers enough energy to propel the vehicle 40 miles per day. Alternatively, for days when the car is not being driven, solar electric production from the charging station is used to offset the electric consumption of our office.

It is a true joy to hear the sound of birds at stoplights. The convenience of not having to stop for gas at "convenience" stores is especially rewarding. Understanding all the costs associated with conventional sources of energy may lead to a wider acceptance of harnessing solar energy in the mountains.

Joshua Bills, solar energy consultant
and engineer, Berea, Kentucky

Transportation is the most difficult Appalachian appropriate technology area to treat. Older transportation modes and networks (footpaths and rural roads) are not appropriate for modern needs; modern road conditions are unsafe for biking and walking; and cars and trucks use too many resources. Any choice we make seems to compromise our growing sense of appropriateness. We realize that improvements in the transportation systems in this region—the return of passenger trains, the creation of bike and pedestrian lanes—will demand coordinated policy efforts and not simply attempts by single individuals.

GENERAL TRAVEL MODES

Appalachian residents have always moved about, using different modes of travel at different times.

Before the coming of Europeans, Native Americans traveled along an integrated network of footpaths and river systems throughout Appalachia, visiting distant tribes and exchanging lightweight goods like shells and craft objects. Many trails became the routes of mountain roads, while others lapsed into disuse.

Foot travel is slow but burns no fossil fuel and requires little equipment beyond sturdy footwear. It is generally less stressful and more enjoyable in areas without motor vehicles. However, attempting to walk in congested urban areas or on rural roadways with no sidewalks can be dangerous, especially on narrow, heavily traveled rural Appalachian roads. At the same time, the region is showing a renewed interest in developing an integrated hiking trail system to attract tourists and capitalize on the popularity of hiking. Currently most transportation funds go toward building, maintaining, and improving modern roads, but there is little provision for footpaths or pedestrian lanes.

Mule- and ox-drawn wagons and horse-drawn buggies were the principal mode of travel in Appalachia in the nineteenth century, and they were used well into the twentieth century. Travel was rather slow and somewhat uncomfortable in rainy or snowy weather, though a driver could nap on the return home on a familiar route. Horses require hay and grain for sustenance, not distant and expensive nonrenewable fossil fuel. But traveling by horse-drawn buggies on roads with fast-moving vehicles, as the Amish do, is risky. In modern America, horse travel is not appropriate from a safety standpoint. As in the case of foot travel, horses and horse trails should be regarded as a viable form of tourist recreation.

For several years in Washington, D.C., I commuted by bicycle 3 miles from a residence on Capitol Hill to midtown, but spills and narrow misses by a commuter bus told me to stop stretching my luck. Bikes are good for sunny weather, light congestion, and leisurely exercise, but they offer even less protection than the more visible buggy. Towns designed like Davis, California, give us hope that many urban places can promote safe biking. Americans experience culture shock when visiting Amsterdam during commuter hours; there are no automobile fumes, only the low swish of passing bikes; life seems so much more ordered and less stressful. However, it seems unlikely that the congested, narrow roads of Appalachia will be made safe for biking given all the interests competing for the transportation dollar. Bike trails are a better option than biking on the average Appalachian roadway, which often have no shoulders. ASPI staff member Russell Parms was killed in 1999 while riding a bike as his ordinary means of travel. Given current conditions in Appalachia, cycling is only appropriate where road shoulders are wide enough to allow for safe travel in our region.

GUIDELINES FOR APPROPRIATE AUTOMOBILE USE

Most Appalachians walk at times and bike on rarer occasions, but they need motorized vehicles to get to work or school or to buy groceries and supplies. Given current conditions, it is appropriate to use the automobile as a source of transportation. However, appropriate use consists in using public transportation when available, reducing travel when possible, and using cars efficiently when auto travel is necessary.

Use Public Transportation

My parents spoke of trips in the 1910s and 1920s in rural Kentucky; they could board local trains and go visit friends and relatives virtually anywhere in the more populated parts of that state and elsewhere. The demise of Amtrak train service in the region has been a great loss for those who are too young or too old to drive, or who do not have access to a car or truck. Often the poor feel the brunt of the lack of public transportation, whether train or bus. That absence is most deeply felt in Appalachia. Concerted group efforts to improve public transportation are very appropriate. Scarcity of public transportation, a result of tighter government budgets, leads to more private vehicles and generally more fossil fuel use.

Reduce Travel

We admire people who take vacations near home or live near their workplace and apart from the motorized modern culture. The trend to work near home is traditionally Appalachian in character, for subsistence farming and craft work kept people from excessive travel. Some modern office workers live in larger cities with convenient public transportation, and some people live in rural areas where they still subsist in tightly bound communities on mixed agriculture. The Amish generally live in productive non-Appalachian farming areas of Pennsylvania and the Midwest.

As of this writing, well over 1.5 million people in both urban and rural areas work from home offices, using their computers, e-mail, fax machines, cell phones, and other electronic devices, and shunning stressful commuter travel. These people prefer the relaxed atmosphere of home over the congestion and stress of workday commutes. More and more find that modern means of communication make associates as available as if they were in the next room; Internet search engines allow access to an extensive information library; and traveling electronically is definitely good energy conservation practice. Even some skills training and college courses can be taken at home.

Nothing would cut local automobile travel more than curbing unneeded trips and doubling up on those needed. Admittedly, traveling alone can serve as a good time for reflection, but much could be done with the minimal inconvenience of dropping others where they want to go. Carpooling with family members, next-door neighbors, fellow church members, schoolmates, and workers taking the same commuter route could have a significant impact on fuel use.

Carpooling offers more than fuel savings; it is also an opportunity to share with others, to keep alert, and to spend social time with like-minded folks. The thoughtfulness of asking others if they need anything when we make a certain trip is part of this trend. Planning to pick up groceries or dry cleaning when going to the doctor or to work is a fuel saver. On longer trips "triangulating"—that is, going to two distant destinations in one round trip, as I and ASPI's environmental team try to do—can often save hundreds of miles. People who live farther from an airport, public transportation, or shopping learn to plan well ahead, or risk running out of time or necessities. Efficiency involves planning.

Adopt Efficient Measures

Some transportation is a must in a civilized society, and modern means of transport are regarded as necessary. Accessible transportation breaks the curse of isolation, improves social life, allows families to bond, offers untold educational opportunities, and gives people a chance to send necessities rapidly to needy parts of the globe. Most people recognize the negative aspects of congestion, air pollution, consumption of nonrenewable resources, auto accidents, and rising insurance fees. For both better and worse, most of us in this region are caught in the private auto culture and must make the best of it.

Because all modern modes of transportation are major energy users (an estimated one quarter of total nonrenewable energy powers ships, railroads, planes, cars, etc.), any reduction in travel or increase in fuel efficiency will impact the environment immensely through reduced air pollution and resource depletion. Some of these changes are more appropriate than others.

Drive efficiently. Faster driving generally means wasted fuel. We increase auto mileage efficiency by reducing speed and use of air-conditioning. Traffic jams are an extreme waste of people's precious time and petroleum, and they generate a deplorable amount of air pollution. Some of these rush-hour tie-ups can be avoided by scheduling work to arrive earlier and leave earlier. Fast starts and excessive braking are inefficient as well as somewhat reckless habits.

Maintain the vehicle. The car manual says it all. Keep tires properly inflated, change oil at regular intervals, replace timing belts, spark plugs, and air and gasoline filters, and so on. Check tires for excessive wear, and rotate them regularly. A competent mechanic is like a good doctor reminding us of needs we often overlook.

Replace inefficient vehicles. Among transportation alternatives now gaining favor again are more-efficient petroleum-fueled internal combustion engines and automotive-powered alternatives to internal combustion engines. All have some merit from an environmental standpoint, but some are longer-range solutions whose implementation could require new policies and commitment by our nation and the world.

Use alternative fuels. Non-petroleum-based fuels are also regaining favor. Biodiesel fuel is an alternative refined diesel fuel. Vegetable oil, or filtered cooking oil residue from restaurants, can be used with slight processing as a recycled product to run vehicles. This by-product was formerly used in animal feeds, but such use is no longer in favor. The fuel source is limited, but advocates say it is a good use for a waste material.

Compressed natural gas is cleaner burning than petroleum products and a relatively safe fuel, but supplies are only as extensive as are our natural gas reserves. Vehicles using this alternative fuel require specific filling station networks not yet in place.

Ethanol can be made from liquid-fuel-based agricultural production and cellulosic feedstocks such as biomass wastes, corn, native fast-growing plants, and short-rotation woody crops like poplar. Currently there is considerable controversy over the net energy gain in transport fuel from various of these sources.

When traditional internal combustion engines are abandoned, the hydrogen fuel cell is a most enticing prospect, since the generating units are decentralized and the only emission is water vapor. A hydrogen economy will require building a sizable manufacturing infrastructure and persuading the public that hydrogen can be handled with relative safety, while making people fully aware of hydrogen's potential dangers. The shift to hydrogen, which will quite possibly occur at least for land transport, will take an immense policy shift on the part of our fuel industry.

Consider hybrid gasoline-electric vehicles. Hybrid cars are excellent for reducing air pollution. They are a preferred option when a new car is needed and can be afforded; they are far better than SUVs, which blur the distinction between cars and trucks in regard to fuel efficiency and whose popularity led in the 1990s to stalled efficiency improvements. Manufacturing accounts for one-fifth of the resources used by a vehicle over its lifetime, but the possibility of sizable fuel and money savings could warrant replacement of gas guzzlers when possible. However, many Appalachian people cannot afford new vehicles, whether hybrid or conventional fossil-fueled. One option is to acquire used nonhybrid cars that were proved by the previous owner to be fuel efficient (e.g, Honda Civic, Toyota Corolla, Ford Escort, or General Motors Saturn).

Hybrid vehicles use electric and gasoline-fueled internal combustion to reduce fuel use and increase efficiency; the rapid rise in their popularity has been astounding. The Honda Insight emerged in the medium price range in 1999 (68 miles per gallon [mpg] on the highway and 61 mpg in city driving). The Insight was followed by the Toyota Prius (52 mpg highway, 45 mpg city) in 2000 and the Honda Civic (51 mpg highway, 45 mpg city) in 2002. In 2004 the Toyota Prius II and the Honda Accord appeared along with the Ford Escape SUV, with three other SUVs following in 2005 (the Lexus 440h, Toyota Highlander, and Mercury Mariner). Three more cars appeared in 2006: the Nissan Altima, Toyota Camry, and Hyundai Accent (the last under $20,000). Federal tax credits are available on the first sixty thousand hybrid vehicles sold by each company based on fuel economy performance.[1]

THE ASPI SOLAR ELECTRIC CAR

Since the automobile is needed in modern-day Appalachia, we ought to look for the most appropriate form of transportation given current conditions. ASPI's solar electric car (figure 24.1) is not run only when the sun is shining (which would be an extremely light, compact vehicle). We are speaking of a solar-powered electric car with batteries that are charged at a charging station consisting of an array of photovoltaic cells on a building.

Here's our case for the solar-powered electric car:

It does not normally use nonrenewable resources (petroleum or natural gas) but runs off energy from the sun. Since the car is electric, its batteries could be recharged by utility-generated electricity, but an extra effort must be made to furnish a solar system for charging the batteries.

It can be easily made by converting existing vehicles with resources already in place with the help of knowledgeable mechanics.

1. David Friedman et al., "Shopping for a Hybrid?" *Catalyst,* Fall 2005, 7–8.

Fig. 24.1. ASPI's solar car is demonstrated by builder Joshua Bills.

It operates more cheaply than gasoline-fueled vehicles. It may or may not have as much pickup from standstill as conventional cars, but it has essentially the same ability to climb hills and travel at recommended speeds.

It is excellent for local trips (within 20 miles), which make up over half of those undertaken.

It constantly reminds us that we need to move beyond the nonrenewable-fuel economy and is part of a transitional phase to a new economy.

The solar-powered electric car serves as an introduction to the solar age and has a promotional and an educational mission to a public that does not regard appropriateness as of much value.

Electric cars were present at the dawn of the automotive era in the early twentieth century, but history and corporate policy turned people's attention to the internal combustion engine and petroleum fuels, which seemed in the early 1900s to be a limitless resource. Thus we have come to witness massive air pollution and drastic depletion of petroleum resources. The time for the solar electric car has arrived.

Recharging car batteries from fossil-fuel- or nuclear-fuel-based power plants does not save resources, since two-thirds of such energy is lost in generation and transmission of the electricity to the point of use. Far less energy loss occurs in refining gasoline or diesel for internal combustion engines. However, resource savings could occur if wind generators, photovoltaics, ecologically sensitive microhydroelectric power systems, and fuel cells are the sources of the electricity for the electric car.

A solar electric vehicle offers numerous advantages. It is created from a used and recycled vehicle, which is no longer powered by expensive nonrenewable fuel. It is extremely quiet, requires no regular engine maintenance, and is nonpolluting. At traffic lights or when the driver's foot is off the gas pedal, no fuel is required, as in internal combustion engines. There are no spark plugs to change, no tune-ups, no motor oil, no antifreeze, and no radiator coolant. An added advantage that is often overlooked is that the electric car can give an instant assessment of driving efficiency by telling the exact amount of energy used at a given moment.

Many would regard the range of such a solar electric car, 80 miles, as extremely limiting. They forget that a very high percentage of all trips are less than this distance. If solar charging stations were available at homes and workplaces, most commuters could be satisfied with these range limitations. Likewise, most trips to grocery stores and school are well within the solar car's range.

The solar electric car is not without some disadvantages, however. It has its highest torque at start-up and thus could squeal tires, but its pickup at higher speeds is less than that of conventional cars. It takes steep hills a little slower than its conventional counterpart. The heavy weight of the batteries may require an improved suspension system in lighter-weight vehicles converted to solar electric use. Trunk room is lost to battery storage, though about half the batteries are up front in the space formerly occupied by the internal combustion engine. The car's range could be reduced by up to 20 percent if a heater or air conditioner is run off the batteries. The sheer number of batteries means that some additional battery maintenance is required over that of a standard vehicle.

A 2,000-pound vehicle (preferably a light pickup truck) is best suited for conversion, although a Dodge Colt donated to ASPI, which weighs 2,150 pounds, proved quite satisfactory. Minimizing starting weight is extremely important to maximize the range on the twenty lead-acid batteries, which in this case weigh 800 pounds; the batteries have an eight-to-ten-year lifetime.

The conversion takes time, semiprofessional skills, and some garage equipment. The how-to manuals are highly detailed, but one has to have the auto lift and tools for the job, proper mechanical skills, and patience. Gutting the internal combustion vehicle involves removing many of the internal working parts and marking and storing those to be reused; the engine block, gas tank, ventilation system, and radiator can be sold. It is also critical to use the factory service manual for the vehicle and especially Michael P. Browne and Shari Prange's book *Convert It!* (see Resources).

The ASPI solar charging station (see figure 24.2) was built after the vehicle was converted, and for a period the car had to be charged via the utility grid. After the completion of the charging station, the entire investment came to $9,200. In a state with the nation's lowest electric rates (5.2 cents per kilowatt hour), recouping costs is more difficult than in states with higher rates, as it takes more years of solar electricity to make back expenses. However, the charging station was tied into the utility system, and much excess energy from the charging station is now used for the office where it is located. More than fuel savings are involved. Thus it is difficult to calculate payback time considering

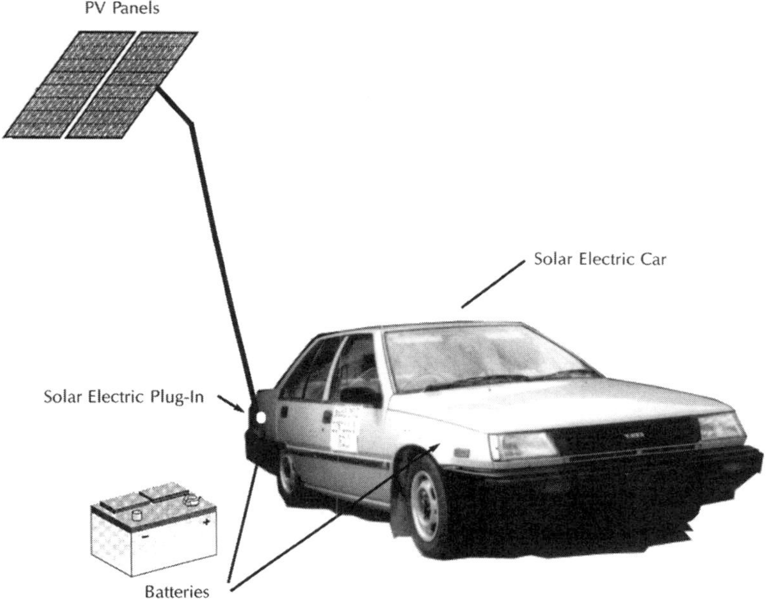

Fig. 24.2. ASPI's solar electric charging station

educational and promotional advantages, electric feedback to the system, delayed car repair costs, and fuel savings.

CHAPTER 25

Composting and Vermicomposting

One of the easiest and most enjoyable ways to adopt an earth-friendly lifestyle is to recycle organic materials using earthworms. Called vermicomposting, earthworms will eat your food scraps, crop and yard residues, and animal manure and convert them into rich humus that is beneficial for your soil and plants. Vermicompost has a magical effect on plants, causing them to germinate quicker, grow taller, produce more leaves, fruit, flowers or vegetables, and develop deeper roots. Vermicompost also suppresses plant diseases, parasitic nematodes, and insects and mites that attack plants. People all over the world are vermicomposting as a waste management strategy, or to make a soil amendment and plant growth enhancer, or as a sustainable business that accomplishes all of this. To find out about vermicomposting, go to http//www.hae.ncsu.edu/topic/vermicomposting/vermiculture/.

Rhonda L. Sherman, extension solid waste specialist,
biological and agricultural engineering,
North Carolina State University, Raleigh

Composting is the process of recycling discarded organic materials in a natural manner so that they will return quickly to humus, which can be used to assist plant growth. Compost, a product of the action of microorganisms on organic matter, is a dark, friable material similar to the dark portion of untilled soil. The process of creating compost has been known for thousands of years;

even the Roman Cato realized that compost materials enhance the fertility of soil. Nature's resurrection cycle is constantly composting organic "wastes" and returning them to living things. A good example with which all are familiar is the forest leaves decomposing in winter and turning into humus for spring plant and forest growth.

ADVANTAGES OF COMPOSTING

If nature composts somewhat automatically, why mess with nature? Nature sometimes acts more slowly and under different conditions than we experience in our "developed" environment. While composting is truly Earth-friendly, we must manage, hasten, and enhance the process involving carbonaceous materials by regulating, either manually or mechanically, the appropriate mixture of air, water, bacteria, and other helpful organisms.

Domestic composting has many advantages: We take responsibility for our own waste and do this in our own backyard, not by sending it out through agencies to distant landfills or other disposal sites. Composting allows us to participate in the natural biological cycle and to truly touch the soil. It avoids our paying fees for waste hauling and disposal. It allows us to put good nutrients right back into the nearby soil without being bothered by odor or vermin. It offers neighbors a valuable lesson in what they can do to improve their own land. And it affords an opportunity to live off the land and thus enter into the spiritual rhythms of place and season.

Over and above what composting does for us as composting agents is what it does for the soil itself. Compost is a more adequate replacement for synthetic fertilizers, which can actually retard or kill beneficial microorganisms needed to keep the soil vital. Composted organic matter allows more oxygen to circulate through the soil, holds moisture and nutrients, and releases these nutrients slowly through natural processes. It helps bring soil to a spongelike condition and allows the excess moisture to drain away from plant roots. The compost helps buffer the soil and keeps the pH near the neutral range, which is best for most plant growth. It helps maintain organic matter in the soil at about the optimum level of approximately 5 percent and reduces the need to import manure or leaf mold. Finally, the compost darkens the soil, helping it to absorb heat from sunlight and extending the growing season.

The first part of this chapter deals with domestic composting as practiced in many parts of the United States, including the use of hotbeds and other home composting methods. The second part deals with a new but equally inviting Earth-friendly technique of dealing with wastes, *vermicomposting,* a term derived from the Latin word for worm, *vermis.* These earthy creatures received Charles Darwin's accolade, "It may be doubted whether there are many other animals which have played so important a part in the history of the world as have these lowly organized creatures."

THE ART OF COMPOSTING

Some basic questions that generally arise about composting are answered here. (Also see chapter 26.)

> *What should be composted?* While domestic waste streams differ markedly, the Environmental Protection Agency estimates that about two-thirds of home-generated wastes could be candidates for composting. These include food scraps such as eggshells, fruit peels, vegetable scraps, coffee grounds, and stale food (but greasy food and meat scraps should go to pets and wildlife); garden and yard wastes such as grass clippings, weeds (except highly invasive plants such as mugwort [*Artemesia vulgaris*]), straw, manure, and leaves; some forest materials, such as reasonable amounts of sawdust, wood chips, and small twigs (larger woody components are good kindling for fireplaces or woodstoves); and certain paper wastes such as soggy

paper towels (glazed or slick paper, newsprint, and office paper should be recycled). Variety in the compost bin is better than too much of a single waste item. Sawdust is often regarded as a waste product of the forestry industry. Actually, it has considerable promise as a weed suppressant when used on garden paths. The sawdust path decomposes in a year with the aid of in-place microorganisms and friendly earthworms and can be a soil amendment for the following growing season in the adjacent garden beds.

Who can compost? Every able-bodied resident. Our nation needs concerned citizens who take responsibility for their discarded organic material.

When can composting be done? At all times. During bad weather, a small lidded compost container covered with soil could be used to collect materials until the weather improves. At each use, it is necessary to remove the soil covering (a small shovel is ideal for the job), open the container, and re-cover it. If not frozen, the compost pile can be augmented even in cold weather.

Where should composting be done? The composting location should be convenient to the user and generally near where most waste is generated. Visitors should not be kept away from the area; if properly maintained, the composting area is a mark of ecological stewardship. (ASPI's composting bins are shown in figure 25.1.) Backyards have convenient shady spots that are ideal. However, if trees are too close, they will send roots into the compost and sap the nutrients. A plastic bottom liner in the compost pit will minimize this loss.

The composting process is quite easy; the challenge is how fast one will produce the compost. The more frequently the compost pile is turned, the faster the process. The best container for composting is a bin made from discarded pallets or waste pine

Fig. 25.1. ASPI's composting bins at the shady border of a small garden plot

slats that allow air to flow easily through the pile. Although pressure-treated lumber will last much longer, it should not be used, to avoid contaminating the compost with chemicals found in the preservatives. A 4 foot by 4 foot by 4 foot compost heap is ideal. Often people line up three bins in a row, one with almost complete compost, one for material at a medium stage of development, and one for newer waste (see figure 25.2). Bins in a shady, well-drained location could have a trench dug around them, as well as another trench down the middle of each bin's floor to improve aeration. Expensive garden choppers, turners, and mixers are for sale but not needed, for composting can be good, steady, medium-effort exercise. All that is needed is a good turning fork.

The ideal in a compost pile is to have organic materials dispersed and thoroughly mixed. Soil is important both as floor and interspersed as layers in the pile. If possible, more small branches, coarse weeds, and other materials that will create air spaces should be near the bottom so that airflow will be

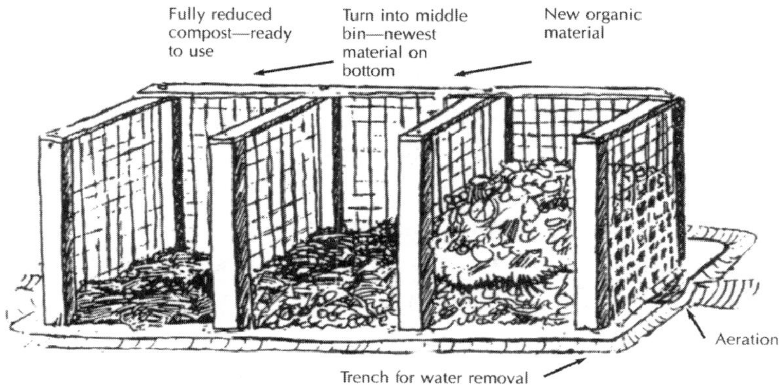

Fig. 25.2. A tripartite composting bin

enhanced. With the proper size (minimum of a cubic yard), sufficient air, nitrogen materials, and moisture, the pile will naturally heat up to about 140 degrees Fahrenheit. While weed seeds will be killed at that temperature over time, so will some or all friendly worms present. At times manure or urine will have to be added to keep a nitrogen-carbon balance of at least 30 parts carbon to 1 part nitrogen. Piles should be moistened occasionally but not soaked.

An alternative composting device is the Indore Composting Pit, a 4-by-6-foot pit about 1 to 1.5 feet deep. On a 1-foot-deep layer of brush is placed 6 inches of green matter (weeds, leaves, or vegetables and fruit scraps from the kitchen), followed by a 2-inch layer of manure, topped with a layer of topsoil and powdered lime. All layers except the brush are repeated in the order given. The resulting pile is regularly watered moderately.

BioActivator, a commercial product sold by Necessary Trading Company, contains beneficial microbes that speed decomposition of organic materials into humus and should be added at about 0.5 pounds per ton of material. This concentrated activator inoculates the soil with beneficial bacteria and tends to reduce any stray odor as well. Some people add mineral supplements to meet the organic gardening needs of their locality (which can be determined by a soil test by the county extension agent) when they start composting.

Protective netting or fencing may be required to keep raccoons, possums, or other neighborhood varmints out of the compost bins. Normally nonmeat kitchen scraps do not attract pests, but cheese and grain products attract rats and mice. Some compost users enclose the bin with ¼-inch rat screen or other protective materials to discourage critters. Our experience is that this protection is generally not needed if only vegetative materials are placed in the composting bins.

It is best to apply the compost to the garden immediately after the process ends. A turning fork is used as a compost spreader when dealing with small composting areas. Some people sift the compost through a screen so only fine particles are put in the garden. Larger, incompletely composted particles can be buried in the garden and allowed to continue composting there or be returned to the composting bins. When compost is spread, the best way to save nutrients is to bury it under soil. If the compost is to be used indoors, it should be pasteurized or sterilized by being placed in a preheated oven at 200 degrees Fahrenheit and kept there for about a half hour.

Hotbeds combine composting of fresh manure covered with soil layers with early plant growing in winter, utilizing the heat from the decomposing manure piles. With this method, plants grow well without using expensive fuels to warm the late winter greenhouse. My great uncle, Louis Burke, was a highly successful gardener near Lexington, Kentucky. In his journals (faithfully kept for over sixty years) the entry for January 2, 1922, called for making fourteen hotbeds that day using nearby horse manure. For him and other gardeners, the heat of the bed did the work of starting the plants (cabbage, broccoli, sweet potatoes, tomatoes, peppers), which would be ready to set out into the gardens by early spring. Hotbed composting allowed the plants to sprout and mature using natural means and in very healthy circumstances. The manure used was composted sufficiently to be applied to the garden immediately after the hotbed materials were transplanted.

Our hotbeds at home were made by putting a layer of cow or horse manure (chicken manure is too hot for hotbed composting) about 6 inches deep on bare ground and adding several inches of topsoil. A series of 8-foot-long, 2-by-12-inch portable wooden frames was set on the manure bed, yielding adjacent 4-by-8-foot beds. These were banked with a plentiful supply of manure up to the level of the slightly inclined frame windows. The vegetable seed layer was thus surrounded by warm composting manure. Our glazing was single-pane windows, but double-pane windows would offer twice the efficiency by conserving heat at night. On warm days the windows were placed ajar for airflow and to allow excess moisture and heat to escape. Some plant watering was required during the lengthy growing season.

VERMICOMPOSTING

Vermicomposting is a subset of composting in which certain species of earthworms are used to enhance waste conversion; vermicomposting produces a very high-grade end-product of worm castings or worm manure. Vermicomposting is a tremendous opportunity for youth and adults to work together in an ideal community project. Rhonda Sherman, who is quoted at the beginning of this chapter, is directing research in the region on this exciting subject.

Vermicomposting is a mesophilic process; that is, worms and microorganisms interact at moderate (self-generated) temperatures (50–90 degrees Fahrenheit), which are lower than temperatures generated inside normal compost piles. The decomposition is rapid, and the castings are of a high quality because they are abundant in microbial activity and plant-growth regulators.

Vermicomposting is not the same as vermiculture, which involves breeding and growing earthworms for fish bait and other markets. Vermicomposting stems from the work of Clive

Edwards of Ohio State University, who over the past two decades has introduced the research of the Rothamsted Experimental Station of the United Kingdom into the United States. In the UK earthworm castings are produced, and through a continuous flow, the indoor operation is segregated from the activity of the worms working in the waste feedstock. Edwards focuses on assessing the effects of vermicompost on plant growth both in artificial containers and in the open soil medium. Plant growth has increased. We in Appalachia can learn from this work, as we do with respect to other technologies.

Organic feedstock is continually added in 1-inch-deep batches to the surface of the composting worm bed, made of trays or racks (see Resources). Feeding must be controlled for fear of overheating (above 100 degrees Fahrenheit) as the feedstock generates heat, which would kill the worms. Vermicomposting does not work well in regular deep composting piles, where temperatures get too high; instead, vermicomposting beds cover large horizontal surfaces in temperate climates where lower temperatures can be maintained.

While indoor operations, even on a domestic level, allow control of both temperature and moisture content, large-scale outdoor operations cover several acres, and operators seek to locate them near feedstock sources such as recycling centers, food processors, beverage producers, and cotton mills.

The Vernalis, California, operation of Pacific Landscape Supply south of Stockton has 70 acres of beds and uses short-fiber waste from a nearby cardboard-recycling facility. However, most American vermicomposting facilities use herbivorous-animal manure and tend to be located near large dairies where the manure is generated. Rhonda Sherman and fellow researchers at North Carolina State University are attempting to extend vermicomposting to chicken- and hog-breeding areas, which generate enormous amounts of manure and can contaminate groundwater and create major pollution.

Many people, even some organic growers and progressive nursery operators, are not familiar with earthworm castings.

They know that certain commercial fertilizers can be detrimental to microbial life in the soil, but they do not realize that this nutrient-rich vermicomposting product is one alternative now commercially available. The Living Soil Company retails small packages of earthworm castings, bottles of casting tea, and casting tea bags for use in gardening.

Vermicomposting is being adopted in India for waste management; in China earthworm castings have been used in medicine for two millennia. Cuba has more recently been using vermicomposting methods to replace interrupted fertilizer supplies from the defunct USSR. Vermicomposting efforts are also being undertaken in Australia with grape wastes from winemaking, and in Mexico vermicompost is used as fertilizer in coffee production. Global applications in waste-prolific and fertilizer-short areas are immense. Market possibilities are expected to extend in this country to grocery chains, nurseries, home improvement centers, and flower shops. Vermicomposting tea could be marketed as well to orchard, nursery, landscape, and greenhouse operators, and community co-ops could find this as profitable as the thousands of small shops that sell fishing bait in our region.

The vermicomposting product is a high-value recycled material that can be used for soil amendment. The process is faster than standard composting, produces fewer odors, and results in a superior product. It is somewhat labor intensive without mechanized equipment and thus is ideal for utilizing a semiskilled labor force in developing countries. The product is proving to be a valuable component in poor soils but even more so in container growing and greenhouses, in leaf sprays, and for treatment of rootstock in vineyards. A suggested two-part program for treatment of poor soil would include treatment with traditional compost to bring soil into productivity and then a vermicomposting treatment for still better production. The vermicomposting product also has insect-repellent properties, which makes the material ideal for organic gardening.

However, the vermicomposting process is not without some difficulties. It requires more surface area; it is susceptible to heat, high levels of salt, high ammonia levels, and anaerobic conditions. In addition, some potential feedstocks can be toxic to earthworms. Generally only highly pigmented varieties of worms can be used. It has been said that the number of worms will, under optimal conditions, double in ninety days; this may be overly optimistic, but earthworms do multiply very rapidly.

CHAPTER 26

Composting Toilets

Several years ago the folks at ASPI, as part of their U.S. Environmental Protection Agency Small Grants Program, installed a compost toilet here at our television studio/office. It has worked beautifully for the past six years and we have become satisfied users and promoters. We hear that this is the only compost toilet in a television station in Appalachia. The compost toilet is part of living the way we ought, consuming less resources and solving the waste problems in everyone's own backyard. Thus WOBZ-TV continues to promote friendly technologies such as compost toilets on its shows, which are viewed throughout much of southeastern Kentucky.

Joey Kesler, director, WOBZ-TV,
London, Kentucky

Appalachia is home to one appropriate technology practice that has immense potential in all parts of the world, especially in sparsely populated areas where sewer systems are quite expensive and in areas where leach field percolation is difficult or impossible.

Compost toilets process human waste into a nutrient-rich fertilizer for use as a soil amendment. Risks associated with water-borne waste disposal, including contamination of ground and surface waters and the spread of disease-causing bacteria, are virtually eliminated by the aerobic decomposition of waste material in the closed-container composting toilet, where temperatures, oxygen, moisture, and carbon-nitrogen ratios are properly controlled.

The composting toilet was developed in Sweden in the 1930s. The extremely rocky ground in the Swedish countryside made installation of centralized sewer treatment systems or septic tanks very difficult. Thus there was a need for a safe way to collect human waste on-site and process it in a way that mimicked the process of natural decomposition.

Disposal of human waste was not a problem for hunter-gatherers. With dispersed populations, wastes deposited among leaves in a forest quickly disappeared through aerobic biological action—nature's way to recycle nutrients. As populations increased, wastes needed to be collected to avoid water contamination and the spread of disease, so latrines were developed, with their attendant unpleasant odors and accompanying varmints. In latrines, waste underwent a slow anaerobic process with some methane generation. Generally the latrine pit was filled in with soil after use and the deposited material was unavailable as a fertilizer.

As populations increased further, more sanitary-minded inhabitants tried to flush waste out of the urban areas by the use of water, sending unprocessed wastes to cesspools in less-inhabited or rural areas. Thomas Crapper outfitted London with such a sewer system over two centuries ago, and similar systems were used to some degree as far back as the height of the Roman Empire. However popular these systems were, they were not totally ecologically sound in concept nor much appreciated by rural families forced to endure the stench of the waste.

Treatment systems replaced unhealthy and unsafe cesspools, but at enormous economic and environmental cost to communities worldwide. The contaminated wastewater mixture has to be cleaned up for reuse, and this is quite complex. Less-expensive individual septic tanks are not perfect, because either groundwater becomes contaminated or, in areas where the soils have high clay content, little percolation occurs and additional sites for septic systems must be identified. With increasing water shortages and declining water quality, the compost toilet promises to contribute to the solution of waste problems and the elimination of water pollution. The composting toilet is truly nature's way.

HOW THE COMPOSTING TOILET WORKS

Composting wastes are not removed to another place by an aqueous medium as is liquid sewage. Rather, they remain in place, composting for a period of time in a container or vault that has access to a current of air. There the waste is combined with carbonaceous material (wood chips, sawdust, peat moss, chopped leaves, or other loose organic materials). The decomposition process generates carbon dioxide and water vapor, which are vented by natural ventilation or by a fan, which can be solar operated. The carbon-nitrogen balance is maintained, and friendly bacteria do the rest, provided the temperature is above freezing. Since water makes up about 75 percent of the half pound of feces and about 94 percent of the two pounds of urine excreted by an average person per day, residual solid matter is quite low. What comes as a surprise is that the final humus volume (3 to 10 gallons per person per year) is so very low.

Pure urine is sterile, unlike fecal wastes, which contain pathogens. Thus the urine with its rich nitrogen content can be shunted off and applied directly to growing plants. This is important, because too much urine in the composting toilet can retard or flood the decomposition process.

The final solid decomposition product (mostly the added sawdust or other carbonaceous materials) is a nitrogen-rich, non-odorous humus, which can be applied as soil amendment to trees, bushes, and berries. If the end product is treated with solar heat to destroy any possible long-lived pathogens (for example, by placing it under a black plastic cover and baking it in direct summer sun for a day), the humus can be directly applied to garden vegetables as well. However, one reported case of food crops being contaminated by fecal coliform bacteria could set back by decades acceptance of composting toilet technology by public health officials.

The number of times the toilet needs to be emptied depends on the amount of use and the size of the container. The humus only needs to be removed every several years if the containment vessels are over 2 cubic meters. Small containers require more frequent disposal.

In addition to these ecological advantages, composting toilets are easy to build and maintain; can be obtained at low to moderate cost, depending on whether one constructs a unit or buys a commercial model; and are inexpensive to operate, as no costly sewage treatment or plumbing repairs are needed. They use very little water. Often domestic water consumption is cut nearly in half, making it more feasible to use cisterns in areas where water is in limited supply. The toilets may be installed indoors or outdoors in a separate building. And with composting toilets, each user takes responsibility for the waste generated, and the end product can be used locally as fertilizer.

MAINTENANCE

All appropriate technology involves some maintenance, but the compost toilet requires little more than the flush toilet. The compost toilet does not have to be designed with a constricted chute as a flush toilet does, and cleaning is reduced because a wide pipe leads down into the container areas.

Before first use, the new compost toilet is inoculated with a bed of material including the friendly bacteria found in ordinary woodland litter and decaying animal manure. This initial inoculation should be sufficient for all future continuous use.

Just as the user flushes a regular toilet, the user adds a cup of carbonaceous material (sawdust, leaves, dried grass clippings, shredded paper) from a nearby bin to the composting toilet. In old-fashioned outhouses, lime or ash was used, but this is not appropriate in the compost toilet, for these alkaline materials reduce the acidity required for proper composting operations.

Just as with a flush toilet, the container must be kept free of foreign objects. Meat scraps and oily products should not be added to the chamber. Most people do not insert fruit or vegetable products, for these attract fruit flies; instead, fruit and vegetable wastes can be added to outdoor compost bins and covered with dirt. Air seals installed at the seat keep flies and other insects from entering the chamber at the seat aperture. Screens should cover the intake air vent and the chimney. As an added precaution, one could add centipedes to the chamber to eat fruit fly larvae regardless of whether plant wastes are put into the toilet; but some people may be squeamish about centipedes.

Excess liquids, including gray water, should be avoided. Gray water can be easily treated in an accompanying constructed wetland (see chapter 16). Some suggest installing an attachment that shunts urine to a separate container; others install separate urinals to reduce urine overload. The collected sterile urine can be diluted to aqueous tea for direct addition to garden vegetables that need nitrogen.

Solid composted material must be removed when necessary and can be added to garden plants after baking to kill pathogens or used untreated around perennials and ornamentals.

TYPES OF COMPOSTERS

Compost toilet units may be divided into two main categories: commercial varieties and do-it-yourself types. The major difference is that prefabricated toilets are generally bulky and more costly (between $1,500 and $6,000) because of heavy packaging and transportation costs. Most of the toilet is a bulky but lightweight fiberglass containment vessel, which can be made as well or better on-site from sturdy materials such as concrete blocks. Compost toilets created from scratch can be produced with very few materials, for as little as $300 plus labor. Labor costs obviously are more than for commercial toilets, but not that much more. Space must be prepared and installation done both for commercial and do-it-yourself toilets. Rural compost users soon realize that the "labor cost" money stays in the region to circulate and encourage other local economic activity, an appropriate technology ideal. Over half of the commercial variety's cost goes to a distant manufacturer.

Compost toilets require space. The larger ones have containers of 2 or more cubic yards and, when insulated, require a chamber or extra portion of a room. However, with proper insulation, these leak-proof chamber portions of the compost toilet can be outside the building. It is always advisable to design and install the compost toilet at the time the building is constructed, even though with some extra effort a toilet can be retrofitted to an existing building. Ideally, the large insulated containment vault is directly below the bathroom.

Commercial Toilets

Among the large number of commercial varieties advertised, some are quite small (less than 1 cubic yard) with limited holding tanks. We do not advocate such small units out of concern that sufficient composting time is unavailable before the unit is full. Nor are we in favor of the commercial varieties that have attached electric heating units; these are neither appropriate technology nor natural decomposing units. The two best-known commercial varieties in our part of the country that are natural composting units with large containment vessels are Scandinavian: the Clivus Multrum, about $6,000; and the Carousel, at $2,600 for the medium size and $3,875 for the large, plus accessories totaling about $900 (2005 prices). Shipping costs will vary depending on distance from the manufacturer. The waste material in the Clivus moves down a chute or series of baffles to a lower end with a hatch to permit removal of material every year or so, depending on use. The Carousel is really a round, swiveling drum with four containment chambers, each of which is turned and emptied after composting for nine months.

Our own experience with both these varieties is quite favorable. Both require very little maintenance and have served well for two decades or more. I observed a disaster in the Ozarks where the manager of a dozen Clivus Multrum units attempted to heat the materials in the vessel with electric heaters, probably to eliminate excess water. The solids vulcanized into a heavy tar that the organization was unable to remove, and thus the entire $60,000 worth of compost toilets was ruined.

Do-It-Yourself Varieties

The concept of the composting toilet is so simple that it is far better and cheaper to custom-make the containment vessel and prepare the space for the specific building. We have used and advocated three do-it-yourself varieties: the Long Branch Passive Solar Composting Toilet (see figure 26.1), which can be produced on-site for $300 to $1,200; the Big Batch Compost Toilet (see figure 26.2), designed by Bob Fairchild of Eastern Kentucky Appropriate Technology near Berea, Kentucky, which can be constructed for $500 to $1,200; and the ASPI Two-Seater, which can be built for $1,000 including a solar fan (see figure 26.3). A modification of the last variety is the One-Seater, which is ideal for use on lightly traveled, seasonally used trails, which allow time for decomposition in the off-season.

The commercial varieties are more costly, but some people do not yet have the confidence or time to do things themselves. Each of the homemade varieties has advantages beyond the size of the container and the ability to build to fit the building. The solar units have lower maintenance costs since they require no electricity. They work quite well and the one installed at the Long Branch office complex (see figure 26.4) is found in an incredibly beautiful setting surrounded by many flowers and vines in the greenhouse; the solar-heated vault does double duty, serving also as a thermal mass for the greenhouse. The batch unit uses two purchased utility carts, so that one can compost while the other is being used. The two-seater requires less space, because one vault is composting while the other is being used.

All varieties must maintain sufficient airflow through the composting material to ensure aerobic action. Grids, baffles, and other methods can be used to keep the humus-forming material

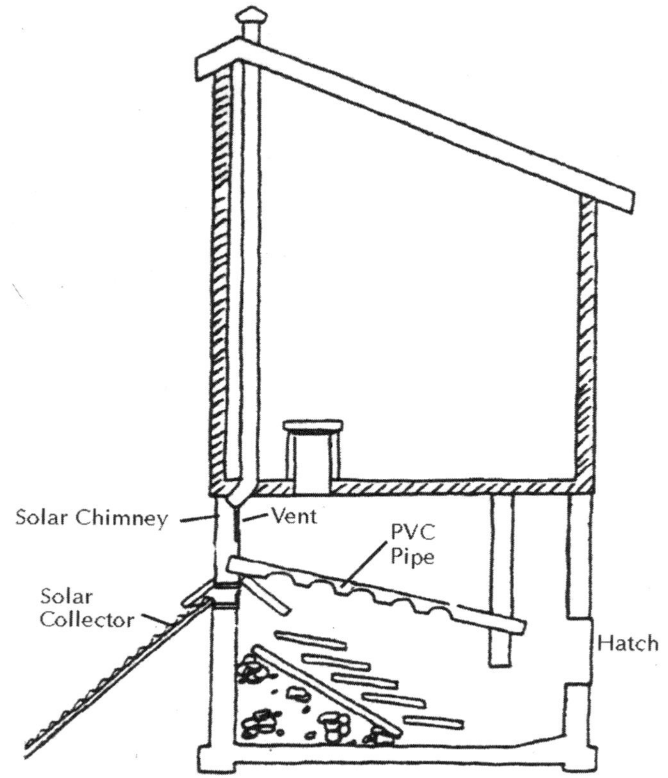

Fig. 26.1. Long Branch Passive Solar Composting Toilet

well aerated and fluffy. Fr. Jack Kieffer, a compost toilet expert at ASPI, suggests an alternative baffling made of rigid corrugated perforated polyethylene pipe about 2 to 4 inches in diameter curving below the composting pile; holes in the bottom of the pipe allow air to enter. The pipe is connected to a standard air outlet going to the outdoors. Often a more powerful fan may be needed to assist in air movement through this baffle.

Small-volume toilets can use a portable containment vessel. One variety uses a 55-gallon drum as vessel held in place by a scissors jack; an airtight fitting at the point of seat attachment

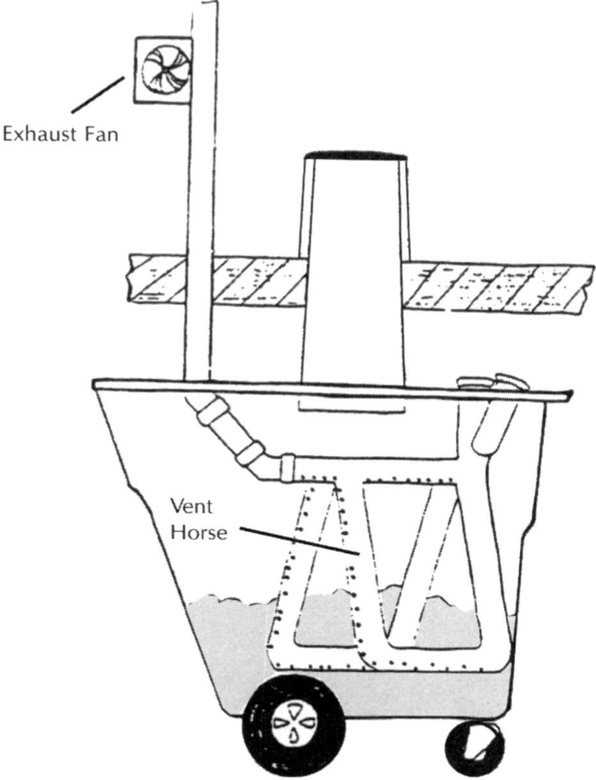

Fig. 26.2. Fairchild Big Batch Compost Toilet

makes it insect proof and allows a flow of air. The vent system is installed in the same manner as described for the above varieties.

ROAD TO ACCEPTANCE

Many regulatory officials side with the construction industry and plumbing professionals, who advocate for treatment plants and sewer lines. Thus there is a major hurdle to clear before compost toilets are accepted by health authorities at the state, county, and

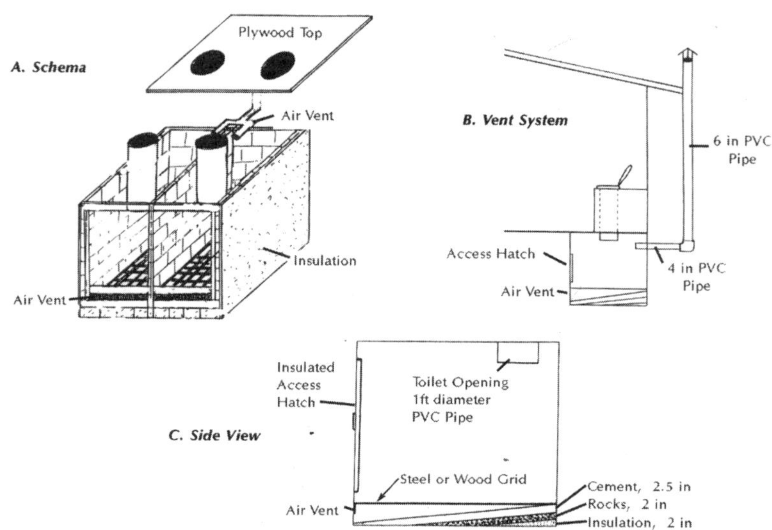

Fig. 26.3. ASPI Two-Seater Composting Toilet

municipal levels. State governments in Kentucky, New Hampshire, Maine, Vermont, Oregon, and Iowa have approved the toilets. Many governmental units approve the toilets on a variance or experimental basis subject to certain restrictions. The patchwork quilt of regulations at all levels makes it impossible to catalog all the rules here. What can be said with confidence is that it isn't safe to assume that rural areas have less-restrictive regulations than urban areas, since many rural agencies try to qualify for EPA construction funds, which have restrictive regulations. Nor does the fact that an area is lower income mean that there will be fewer requirements.

Composting toilets will gain more widespread acceptance when extension services advocate for them, especially in rural and isolated areas.

In some states, compost toilets are installed in backwoods areas, where they work quite well when lightly and infrequently used. However, any compost toilet can be overloaded, and some

Fig. 26.4. Composting toilet inside attached solar greenhouse at Long Branch office complex

urban parks and heavy-use areas have experienced overloading. In areas of extremely high or "spiked" use as on weekends or holidays, we suggest the use of accompanying urinals, which will cut complaints and boost the acceptance of composting toilets. There is no question that all appropriate technology works best with responsible users and can easily be damaged by irresponsible ones.

Some local ordinances require accompanying sewer or septic tank hookups, a requirement that partly defeats the rationale for the compost toilet. Other agencies require French drains in place of septic systems provided systems do not become overloaded. However, a sufficiently large constructed wetland (see chapter

16) meets all the requirements for wastewater disposal and will serve the same purpose.

The compost toilet is not going away. There are too many satisfied customers, and the advantages of the compost toilet are too obvious to users. When the regulatory hurdles are overcome, the compost toilet will be as popular as the well-regarded constructed wetland. Coupling the two is a powerful way to be simultaneously economical, ecological, and resourceful.

CHAPTER 27

Recycled, Salvaged, and Deconstructed Materials

A lot of waste around our homes and workplaces can be recycled. All the materials that we recycle make the county landfill, built at taxpayer expense, last much longer and help us have a healthier environment. . . .

Several of Cumberland County's 14 waste disposal convenience centers, located in communities throughout the county, also accept materials for recycling.

Brochure from Cumberland County, Tennessee, recycling center, named in memory of Irene Power Dickinson (1915–1995)

A critical component of being Earth-friendly is to respect materials and treat them properly. Recycling gives more life to used materials that might otherwise be discarded. But isn't it correct that nothing can be totally discarded on Earth? Affluent people sometimes act as though what they throw away no longer exists, although it may impose burdens on the less affluent who have to deal with it. There is a mind-set that regards recycling as a type of child's play, junk as always existing elsewhere, and waste as a necessity. Actually, from an Earth-friendly standpoint, *recycling* is a necessity, *junk* is evidence of a disordered state of mind, and *waste* results from failure to see the good in resources. We must recycle: we salvage unused stored materials; we see the value when we tear down or are tempted to discard materials. We all need to occasionally review practices that can make us more materials conscious.

RECYCLING MATERIALS

Use what is needed and not more; what you can't use now, reuse later; what you can't reuse later, recycle; what you can't recycle, sell at a yard sale; and for what you can't sell or give away, create a new use. Only as a last resort should any of us send materials to a landfill or incinerator—the final defeat for those who believe that all the Creator made is good. Certainly, some mixed materials (plastics and metal) are difficult to recycle, but Earth-friendly people resist being forced to discard as a general rule. When we become aware of how hard it is to reuse, repair, or recycle a particular product, we have a powerful incentive to avoid purchasing the product or to identify a useful substitute. The very first of the resourceful Rs is *rethink,* then reduce, reuse, repair, or recycle.

America has been home to both good and bad recycling practices. An example of the first is Appalachian patchwork quilts. These traditional bed coverings are made by recycling used garments into highly crafted objects, which bring considerable money (hundreds of dollars each) back to cottage industry people today. A subdivision of that craft is quilted window shades (see chapter 19), which are not only beautiful but also provide home insulation. However, these labor-intensive craft articles may be somewhat too high priced for a larger energy-conscious market. On the negative side is the litter that mars highways and vacant lots. This national problem has been verified by Aluminum Anonymous of Delaware, which sent counters to selected highways in every state. No region is immune, though states that offer refunds on returned bottles appear to have less roadside litter. Proper waste disposal is a national problem.

When I grew up on the farm, we had fewer throwaways, and most of what we couldn't use we passed down to others. Food leftovers, if any, and cooking scraps were added to dishwater and fed as slop to hogs. Scrap iron was recast in my dad's workroom/blacksmith shop into everything from gate hooks and hinges to angle irons and reinforcement for foundations. Scrap wood and yard wastes were composted or burned to make plant beds in late winter, when we burned materials to kill weeds, loosen soil, and add ash for plants. Paper became kindling for starting fires in our coal and wood heaters in home and workshop. Even glass was often crushed and used as a substitute for sand in concrete. That principle of discarding little left only a few items, which we placed as dams in eroded washes in the field or in old sinkholes, which were then covered with topsoil and reclaimed (really a bad practice, for subsurface contamination could occur even if only cans and bottles were buried in karst country, where the soil was largely limestone). Even our scattered stones were gathered, pulverized, and spread on the fields as mineral supplements for pastureland. Materials were used well and not carted outside the farm.

The same lifestyle was exemplified by native Kentuckian Albert Baldwin. Baldwin, who lived near London, Kentucky, on Wolf Creek Lake, said he had never taken a thing to a landfill in his life. His tidy farm of about four acres was a remarkable place. He and his wife Becky raised chickens to consume food scraps, buried his few tin cans as an iron source for his extremely healthy fruit trees, raised all types of vegetables, canned the surplus, reused mason jars year after year, used paper as garden mulch and tinder, and built his own furniture and buildings from local materials. Television stations did shows on his remarkable conservation measures. We were so impressed by his expertise, even though he was unlettered, that we asked Albert to serve as a member of the ASPI board, which he did from 1986 until his death in 1996.

Materials from ASPI have gone for years to a multisort waste-recycling center. We recycle metal (aluminum and others), glass (green, brown, and clear), number 1 and number 2 plastics (PET and HDPE), old newsprint, telephone books, slick paper and catalogs, office-grade white paper, colored office paper, computer paper, mixed paper, cardboard, and rags. We use some items on-site: metal for planters, newsprint and shredded paper for mulch, jugs for greenhouse heat storage, and crushed glass as a sand or aggregate substitute.

Wood wastes from sawmills and processing plants are often regarded as problems. Could they be useful in some way beyond returning them as nutrients to the forest floor (leaves, twigs, slash, stumps, and roots)? Sawdust can be used in composting toilets, along garden paths, and as a soil amendment. Shavings and chips are good for animal bedding; pine and other slabs have multiple building uses, from fences and compost bins to interior and exterior siding and roofing (see chapter 23). Waste pine slabs can be used to border raised beds in the garden; they are not durable but they add to the composted material. Log ends used to border raised beds last longer than pine slabs but are better used as cordwood for mobile-home siding (see chapter 22).

SALVAGED AND DECONSTRUCTED MATERIALS

The mentality of "out of sight, out of mind" affects most of us. Manufacturers take advantage of this adage and are not the champions of the phrase "to reuse is to be Earth-friendly."

Almost all of us can recall wondering why one building or another is being torn down. We note that our country does not have any sympathy for heritage. I opened the newspaper a few years back and was shocked to see a picture of a nearly torn-down rural homestead. It was my great-grandfather John Fister's place, Placentia, where a previous owner had entertained the great Marquis de Lafayette on his return visit to America in the early 1820s. The article said the Bluegrass Historical Trust had failed to save the building, but the workers were salvaging some very nice windows and frames. So much for my own heritage. The principle *save and reuse, do not tear down* seems to be more and more forgotten.

The story of such losses is repeated again and again as old homes, business establishments, and churches are torn down to make way for newer buildings. Occasionally an older church or school is incorporated into a newer (not necessarily better-constructed) one and very little deconstructed material is generated. But many buildings such as orphanages and insane asylums are not part of ongoing enterprises, and they await demolition. Down they come, and high-quality wood and other materials are discarded. Even some roofing and plumbing are of value when torn out. This brings us to a second principle: *respect older structures.*

Neither of these two principles is sacrosanct, for some buildings have decayed so much that they are dangerous and simply are too costly to restore—although sometimes restoration is attempted. Fort Jefferson, off the Florida coast, is being restored at a cost far greater than the original construction cost of this brick and masonry structure. It was never used for war but served as a prison for Dr. Samuel Mudd, who happened to tend to the broken leg of Abraham Lincoln's assassin, John Wilkes Booth. Military observers say that Fort Jefferson was not properly constructed on the sandy soil and had settling problems almost immediately. Is a poorly built structure of the past worth saving, or should it be allowed to become a ruin, perhaps of greater historic value? That is the question being discussed with respect to many Southern antebellum homes, which cannot be easily winterized and have been turned into storage barns.

Deconstruction can produce reusable materials. Building material recyclers have large areas of all sorts of waste materials at bargain prices. At one Nashville building-salvage place where we were filming one of our *Earthhealing* shows, we observed an astute, wealthy-looking gentleman carefully picking over material to find just the right board.

But what about the mountains of old refrigerators and less-usable materials? It takes expertise to keep such materials moving and business thriving, because the best pieces are picked up for a song and the worst are never sold. Moving poorer-grade materials is a challenge. Maybe we need to have a tax subsidy for those who handle and sell such materials. What about a deconstruction tax for each building destroyed or perhaps a disposal tax on all discarded appliances and tires?

At Long Branch Center, coauthor Paul Gallimore has attempted to make a science and art out of imagining what deconstructed materials can be used in all of the center's buildings, seeking ways to turn discarded wood into building materials, especially wood framing and flooring, and older doors and windows and their frames. The center has ladders, porches, stoops, steps, support posts, railings, siding, and trim that are all salvaged from local deconstruction projects.

Many Appalachian creative homesteaders and do-it-yourselfers engage in similar recycling activities. They have ingeniously turned barn siding into interior paneling for medium-priced and upscale homes; they find uses for good concrete blocks and bricks, while allowing mixed debris to be used for parking spaces and roadways.

Cleaning up and sorting deconstructed materials can be labor intensive, but with proper management this work can be beneficial for both the individual and the community as a whole. The trouble is that by sorting through such mountains of discarded materials, one is left with mountains of less-useful materials. Plumbing and electric materials are hard to use because current building code specifications can only be met by new utility products. Some materials are just too jumbled or mixed to sort.

It should also be borne in mind that some deconstructed materials may contain asbestos or other hazardous substances. These pose a risk not only to reusers but also to workers who carry out the deconstruction. Many buildings are basically unsafe at the time demolition occurs, so there are physical hazards as well. Deconstruction comes at a cost.

DOMESTIC SOURCE REDUCTION

The best way to tackle the waste problem is not by recycling but by refraining from making a mess in the first place—source reduction. Counter to our consumer economy is the principle *don't use if you don't need to.* Resource utilization expert Art Purcell, of the Resource Policy Institute of Los Angeles, says that we should focus attention on regional, national, and international source-reduction schemes such as industrial materials substitution and conservation schemes. Too much energy is often given to promoting local or personal source reduction when the focus should be on encouraging creative ideas for broader source reduction.

What can be done with the massive waste stream? How can we change the mentality of people who are addicted to consumer buying habits? (Some people do work to change this mentality; one such effort is shown in figure 27.1). We do not need all that junk (correctly defined before it is bought and not when discarded). We must make source reduction a first principle of environmental resource conservation.

To say wastes are not a problem is simply false. It leads to the assumption that everything is recyclable if we find the person or technique to do it well. Currently, a nonworking computer is given away as though some poor soul is out there craving a computer, even one that does not work. We Americans need to have standardized appliance outer units as in Europe; the appliance is rebuilt and reused when the interior apparatus ceases to function. We ought to replace planned obsolescence with a sounder national resource conservation and materials reuse policy.

Refraining from engaging in the consumer culture is the most drastic means of source reduction, but certainly not the only one. Sometimes a new and innovative item is just what is needed—but not always. People should be encouraged to buy used goods—clothing, books, autos, furniture—when possible. Yard sales are good sources. People can assemble a list of long-

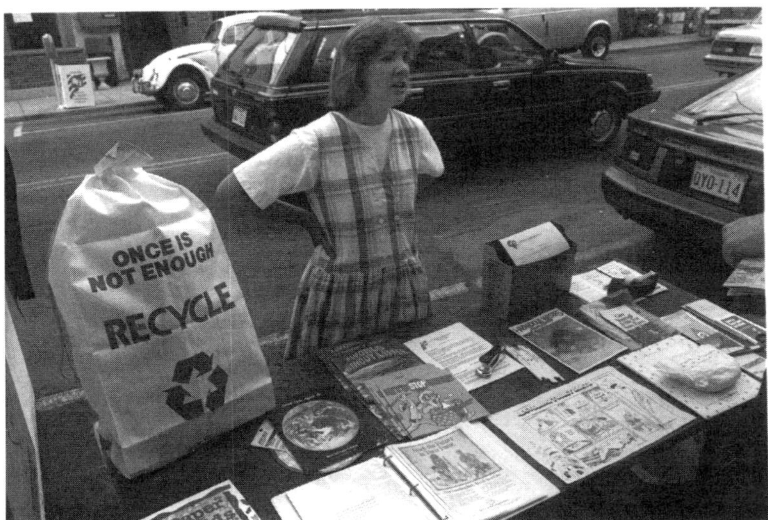

Fig. 27.1. Encouraging recycling in Berea, Kentucky

term items needed (special pliers, rolling pin, etc.). True restraint is to buy only those items on the list, but that seldom happens. Or perhaps the bargain hunter practices some impulse purchasing. Some offer the rather reasonable excuse that the items for sale would otherwise be discarded, and it's better to reuse them than to throw them away. Another rationale for buying is that the item might be needed sometime in the indefinite future. Some are able to frame an item as a gift for another or as a present for someone's birthday.

In addition to refraining from purchasing new products, source reduction can be accomplished by getting rid of our excess. Most of us have items that we don't use but that are hard to part with. Such unused items take up valuable space and require time and money for maintenance. If we are not inclined to sponsor a yard sale, we may find some neighbor willing to sell our items on consignment. Silent auctions for nonprofit groups are another way to pare down our surplus possessions, permitting us at the same time to support a worthy cause.

While it is better to give than to receive, some discretion should be exercised. Do not give away items just to have an excuse to buy more, and don't give away things that are in such poor condition that they are no longer useful to anyone. Low-income people already have a disposal problem of their own.

And our planned-obsolescence, consumerist society has a waste disposal problem as well: the need for more landfills and incinerators. It is not inappropriate to fight such projects, especially when they are targeted for poorer neighborhoods. Refraining from being a NIMBY (someone who says, "Not in my backyard") is hard, but it is needed if we are to heal our Earth. This resistance to socially irresponsible discarding of wastes by or on others can be a step toward getting citizens to reconsider generating excessive amounts of short-lived consumer goods in the first place. That is why tackling our own waste problems at home is such an important first step.

CHAPTER 28

Ponds and Aquaculture

Through my teaching and research at Kentucky State University, I find a growing interest in aquaculture as a source of income for residents in this region. Extensive to intensive fish production systems have been tested and have shown potential success with a diverse group of fish such as catfish, bass, paddlefish, shrimp, and trout. There are some problems associated with raising fish, but they are not insurmountable. We are here to make your life easier, if you choose to roll up your sleeves and get your feet wet. You may wish to visit our website (www.ksuaquaculture.org), join the Kentucky Aquaculture Association (www.kaatoday.org), seek information at www.msstate.edu/dept/srac, or contact us at (502) 597–8103.

Steve Mims, aquaculture research director,
Kentucky State University, Frankfort

Large and small bodies of water are essential for American rural communities. They provide flood control, recreational opportunities, potable water reserves, dependable barge transportation, some hydroelectric power, and aquatic life. For instance, the Tennessee Valley Authority (TVA) lakes, along with private and public lakes, provide these advantages for residents and visitors alike. Larger impounded water bodies, such as the TVA lakes, have certain benefits, even though ecologists point out that such impoundments can have negative effects on watersheds (e.g., lack of fish movement, loss of mussel diversity, and siltation problems). These lakes are very popular tourist attractions, although some naturalists would argue they are not as attractive as natural lakes. We focus here on smaller artificial water bodies typically found in rural areas or near homesteads in Appalachia and other parts of the eastern United States.

RURAL FARM PONDS

The U.S. Department of Agriculture extension service offers assistance to property holders in installing small lakes or ponds on their land. Part of this help is financial and part is advice to ensure that the ponds serve several purposes, primarily that of providing a dependable source of water for livestock.

A well-constructed and maintained pond helps replenish the water table and curb soil erosion by slowing the speed of runoff and contributes to overall watershed flood control. The pond also provides water for fire control, for early garden and flower plants, and for irrigation during droughts. It likewise provides a place for recreation and a source of fish, which can supplement family food needs.

Of almost equal importance is the beauty a good water body adds to the landscape. A private lake or pond is a major asset to any property, for it provides a source of land/water harmony. For the appropriate technologist, this is highly important, for harmony is a sign of good ecology and the desire to keep a place tidy, well maintained, and well ordered for healthy living. The pond is an ideal location for meditating and relaxing and enjoying the surrounding countryside.

Plants at the water's edge add beauty and wholeness to the landscape and also supply natural food and habitat for fish. Care should be taken in choosing plants, because exotic aquatic species such as duckweed and water hyacinth are easily established and can spread beyond their boundaries and become invasive. (Invasives are regarded by many environmentalists as the most serious problem, perhaps second only to global warming.)

Ground-layer plants beneficial to wildlife, including some that are medicinal and/or edible (see table 15.1), include horseradish, swamp milkweed, black mustard, barley, common camas, safflower, common sunflower, bird's-foot trefoil, white lupine, alfalfa, peppermint, spearmint, watercress, American lotus, East Indian lotus, spatterdock, rice, millet, silverweed, yellow dock, Wapato or duck potato, Virginia glasswort, rye, common chickweed, comfrey, strawberry clover, Alsike clover, swamp cranberry, cranberry, water speedwell, and northern wild rice.

Some shrubs, trees, and perennials from 2 to 20 feet in height that can tolerate wet conditions in wetland areas and on pond edges are coyote brush, Jerusalem artichoke, sea buckthorn, damson plum, black currant, American elder, viscid bulrush, broad-leaved cattail, stinging nettle, elderberry, raspberry, blackberry, and highbush blueberry. Suitable shrubs, trees, and perennials that may grow over 50 feet tall are red, black, and white alders; shellbark hickory; white and weeping willow; vine maple; Allegheny chinkapin; sour cherry; common plum; and pear.

Farm ponds or lakes also offer very practical advantages. With the growth of environmental consciousness it has become apparent that they help prevent water pollution by providing a place for livestock to drink other than free-flowing streams. The livestock pond can be equipped with a pipe that provides water to a trough at which the animals may drink without depositing animal wastes directly into the pond water. Lakes are built so that water can flow into them either from surface water or from springs. On our farm in northeastern Kentucky we had several pools, which were walled ponds at places where there were springs and water was collected for cattle watering. For the most part, these succeeded except in the driest of times. We accepted federal money to build a large farm pond in the center of the property, and this settled water-shortage problems for our relatively large herd of beef and dairy cattle.

A major use of ponds and lakes is some form of water sports, which are some of America's favorites. The desire to be near water (over half of Americans live within an hour's drive of a major body of water) is part of being who we are. Eastern Americans share this affinity for water and water sports. How else can we explain the desire for swimming, wading, boating, fishing, water skiing, sunning on the beach, and just looking out over the water? Not all water sports are mutually compatible, so some bodies of water or beaches are reserved for specific purposes, perhaps permanently or at different times. How can youngsters wading in shallow waters compete with nearby speeding Jet Skis and motorboats? Ponds and small lakes can be restricted to non-motorized watercraft, so that even the wildlife will be able to enjoy the tranquillity.

And, of course, farm ponds or lakes can be used to raise fish, whether for personal use or for "catch-your-own" operations. Thus we turn next to aquaculture.

AQUACULTURE

A number of methods of raising fish (aquaculture) are being used at this time in Appalachia. Large artificial hatchery or commercial fish farm operations involve complexes of tanks either outdoors or in greenhouse- or barnlike structures and may cause major environmental degradation similar to that caused by monocultural agriculture. The North Carolina State Fish Hatchery on the grounds of the Pisgah Center for Wildlife Education near Brevard is quite popular for youth who learn about appropriate fish-raising techniques. Most large aquacultural systems are in any case unsuited for home production.

Solar algae tanks, suitable for domestic or community use, developed by the New Alchemy Institute, involve the use of tilapia in naturally fertilized fiberglass tanks that rely on sunshine to grow phytoplankton to feed the fish. Some of these tanks are still used today even though the institute no longer exists as such. Rodale Research Center in Pennsylvania has developed a "home aquaculture unit" using a small swimming pool as holding tank, a water- or air-powered rotary biofilter, and a water clarifier. A plastic dome increases production of warm-water fish and extends the growing season. We suggest this approach for school and small community projects.

Our focus here is on small ponds stocked with native fish and using few outside controls other than external feeding during part of the growth cycle. In many states exotic fish are illegal, yet tales of exotic fish going rampant are becoming more common. "Think native" is a general rule in the eastern United States as elsewhere. If pond water temperatures allow, a variety of fish can be raised in a polycultural system, partly shaded, fed by fresh water, banked by native trees, berries and plants, capable of providing some nutritional needs, and integrated into farm operations.

Different kinds of fish, of course, have different needs. Trout require fresh, clean, cool water (50 to 75 degrees Fahrenheit); adequate tree shading of the incoming stream and the pond itself can help keep water temperatures cooler. (Note that rainbow trout are not native to southern Appalachia and thus brook trout are regarded as the preferred native trout species in North Carolina and elsewhere in the region.) Catfish tolerate warmer temperatures, more silt, and lower oxygen levels than other fish. Both catfish and trout can survive the Appalachian winter but do not put on much weight in cold weather. Catfish are omnivorous and can live on plants, algae, crayfish, and other fish. Light must be optimal for certain fish to feed and reproduce, and sunlight of course varies with season, weather, shading, and condition of the water. Phytoplankton (microscopic algae) flourish with light and may be overabundant in too fertile ponds, whereas the presence of zooplankton (microscopic animals) does not cause a problem.

Both the quantity and type of fish raised are important in successful smaller operations. A polyculture of several varieties is more ecologically balanced than a monoculture. The bass-bluegill combination is conventional in farm ponds with the addition of channel catfish and baitfish (smaller fish eaten by larger carnivorous ones). Other aquatic life like geese, ducks, and amphibians adds to the biodiversity and may help fertilize the pond through their manure. In many areas, some of these waterfowl, especially Canada geese, arrive uninvited and miss their Canadian flight schedules as well, sometimes settling in as permanent guests.

In fish selection, consideration should be given to natural water conditions (temperature, pH, turbidity, etc.). Some fish stir up the sediment and may have to be avoided. Most fish prefer neutral and higher pH levels between 7 and 8.5, which can be obtained by liming. In small operations, adequate dissolved oxygen is attained through the addition of fresh water, surface disturbance, and submerged oxygenating aquatic plants. The quantity of fish to be raised in the pond should be directly tied to the amount of oxygen available. Increasingly, appropriate technologists suggest solar devices for additional aeration during hot, dry weather. Some fish do not tolerate excessive nitrogen compounds resulting from an overload of organic matter, so larger operations install biofilters (made from clam shells, gravel, plastic net, or corrugated fiberglass). Additional fertilization may be needed. The fish obtain calcium through liming, phosphorus from bone meal and rock phosphate, and nitrogen from animal manure and compost. In rare cases, the addition of trace minerals (magnesium, manganese, cobalt) is needed.

Many fish prefer smaller fish for food—and may even eat their own fingerlings, if these are not sheltered. For aquaculturists wishing to provide their own fish feed, a menu of many native aquatic organisms is available. Fathead minnows, golden shiners, and goldfish are commonly grown for bait and can be live feed. If given enough shelter, they will reproduce in most eastern American ponds. Tadpoles and frogs are relished by large-mouth bass. Red wiggler worms are often grown beneath rabbit hutches, to digest rabbit manure; one useful product of this vermiculture is fish feed.

Commercial fish feed (grain, soybeans, fishmeal, and additives) is often added to hasten fish development, but this is expensive. Small-pond aquaculturists generally prefer to concoct their own fish feed with materials close at hand. Quite often the water-borne plants and animals in the pond are sufficient for the stocked fish. But without additional food sources the rate of fish growth is naturally quite slow. Additional food increases growth, and some may want to slow or hasten fish growth for a variety of reasons—for example, to reach a certain size for market or for a festival. An "edible pond bank" can a be source of fish food. The catalpa or fishbait tree grows well at most eastern American pond edges and serves as host for caterpillars, which fall from the trees into the pond, much to the delight of insect-loving trout in late summer. Mulberries are also good fish food. Aquatic plants like cattails, arrowheads, and watercress are part of the total food cycle in many ponds. With some ingenuity a variety of groundnuts and seeds (acorns, chestnuts, hazelnuts, and honey locust pods) can be converted to fish food through hulling, grinding, and mixing. With time fish from the pond become food for the table. Fresh fish is often hard to buy in parts of rural America and is far superior in nutritional value to expensive imported protein sources. A small pond can furnish a family at least one meal per week throughout the year. Because they take up relatively little space, aquacultural ponds are ideal food producers in mountainous areas. At the Long Branch Center, approximately 1,000 pounds of trout per year can be raised in a 2,000-square-foot (1⁄20 acre), 8-foot-deep pond of free-flowing water. The mountain stream that enters the pond flows through a forested watershed and is spring-fed by dozens of high-elevation pristine sources.

Aquaculture offers advantages beyond providing fish for food. Fish themselves are beautiful, and people who feed fish come to appreciate the response of fish to their presence. Fish can keep the pond free of pests such as mosquitoes. Certain types of fish, including channel catfish, are good at removing algae, which is a major problem in water bodies used in home heating and cooling systems (e.g., water source heat pumps) that require relatively clean ponds. Algal blooms may cover ponds in midsummer, especially in those water bodies richly laden with animal wastes. Hardier fish may help keep such ponds clean and

Fig. 28.1. Trout pond at Long Branch

clear. One may conceive of the fish as part of an integrated cycle: pond water is used for irrigation, plants for fish feed, and bottom pond muck and fish manure for fertilizing fields.

PLANNING FOR PONDS AND AQUACULTURE

Because a poorly sited pond is of little benefit, anyone who wants to build a pond should consult with soil conservation experts about the best location. Siting criteria include aesthetic considerations, proximity to dwellings and other buildings, access to a spring-fed source, control of sufficient watershed, and primary uses.

It is important that the pond be located in an area that permits monitoring access to it, to prevent entry by trespassing neighboring youth—a major problem in Appalachia, where commons sometimes include other people's land. Building a pond where it can be easily monitored by the property owner or a willing neighbor may be essential to reduce the establishment of "positive enticement" to injury or death. Steps such as posting No Trespassing signs and putting up fences are generally not sufficient to eliminate the possibility of a lawsuit.

Soil conservation experts can also recommend the best kind of lining to use. Generally a clay lining is the most economical way to stop leaks, and it is far cheaper than using synthetic materials, such as plastic liners, in rocky and karst areas, where sinkholes and leaks occur.

Another consideration in pond construction is watershed control. When property holders control the entire watershed above the impoundment, there should be no threats to water quality from chemical runoff. If the water is fed from urban, suburban, or rural landscapes where water pollutants are not controlled, the lake obviously has limited value in aquaculture. For fish production, we strongly suggest either complete control of the watershed or a formal compact with others within the watershed to refrain from using pesticides and other chemicals on their land to avoid water quality degradation.

The county agent can assist in choosing appropriate Appalachian wetland plants for pond and adjacent pond bog areas. When invasive exotic species such as purple loosestrife are present, they should be eradicated. Suggested native Eastern American plant replacements include golden club, seep rush, grass pink, mountain sweet and purple pitcher plant, meadow sweet, sphagnum moss, and cinnamon fern. Suggested berries include blackberry, raspberry, blueberry, elderberry, and cranberry. Native trees include those listed earlier in this chapter along with black walnut, chokecherry, red maple, mulberry, sycamore, and hazelnut.

As mentioned earlier, allowing livestock free access to a pond will lead to water contamination from animal wastes. This can easily be controlled through proper fencing and piping. The livestock pond can be equipped with a pipe that provides water to a trough at which the animals may drink without depositing animal wastes directly into the pond water.

Year-round free-running streams can have enough oxygen to permit the raising of trout, as at Long Branch's aquacultural pond on the Long Branch of the Big Sandy Mush Creek in the Newfound Mountains. Again, rainbow trout are not native to Southern Appalachia and thus brook trout are preferred.

Fish, like all wildlife, value their freedom of mobility, but they may have to be caged for successful polyculture. Sometimes segregation is required because of preying or aggressive habits of other aquatic or wildlife predators. Obviously, it is easier to harvest the caged fish, but it is not the most entertaining sporting activity. Caging may result in diseases and die-offs, theft, supplemental feeding requirements, and loss from cage damage to the enclosed species, so cages should be used only when truly necessary.

CHAPTER 29

Cisterns and Water Catchments

I loved the taste of Aunt Mary Elizabeth's cistern water. It really beat the heavily chlorinated city water coming from our tap. She and Uncle Albert took great pains to keep the cistern well maintained and that must have contributed much to the water being of such high quality.

Jerry Schumacher, vice president,
Security Bank and Trust, Maysville, Kentucky

Appalachia, like much of eastern rural America, has known the benefits of cisterns from the time of pioneer settlements, even though springs and wells were the principal sources of domestic drinking water. These cisterns, when protected from surface contaminants, provided a dependable source of water, especially during dry times. Today, cisterns remain a source of drinking water for humans and pets and of water for watering small gardens, refilling fish tanks, and washing vehicles. Municipal water is bad for gardens and houseplants.

Cisterns are usually fashioned from rock, masonry, metal, or plastic-enclosed tanks built either aboveground or partly or totally buried. These are generally attached to buildings that serve as the collecting or catchment area. Cisterns had a proven track record long before Jeremiah the prophet was thrown into one. In dry Middle East areas these vessels furnished good water to tide people over from one rainy season to another.

Appalachia is ideal cistern country because it normally has plentiful rainfall in the winter and spring with generally drier summers and falls, when the cistern water can be withdrawn for use. If sufficient protective measures are taken, the water can be potable (suitable for human drinking), but it is often used as soft water for washing, especially hair and delicate clothing, and for watering gardens and scattered plants. Water from cisterns can be a supplemental source for people who depend primarily on ground- or surface water. During dry times punctuated by rain showers, cisterns have a surprising capacity for replenishment provided the collecting surface is large enough.

Cisterns have many advantages, including low unit cost with rapid payback. The amount of initial investment is a known quantity, and building a cistern is far less risky than well-drilling gambles, which often prove quite costly. Cisterns can be used as a water-conservation measure in areas where water is not plentiful. When properly protected, cistern water is soft and without worrisome iron and other minerals found in so much Appalachian groundwater. Naturally soft water is better for people suffering from excess salt, who often forget that artificially softened water can be high in sodium. Cistern water is preferred over chlorinated water for plants and pets, and people find its taste excellent when the cistern is well maintained. A properly sited cistern can also be quite convenient, and it is under the control of the operator, not a distant private or public water agency. If given reasonable attention, the cistern requires little maintenance and repair.

Because cisterns are of limited size, it is imperative not to waste their water. The three-quart, full-faucet-running tooth brushing is out. But this limited water situation can be a teaching opportunity. Cisterns are again attracting intense interest in Appalachian areas where well water has deteriorated due to aquifer fracture caused by blasting operations, where groundwater contains minerals or is contaminated by sewage or hazardous agrichemicals, and where it is too costly to extend water systems to remote settlements. As with all appropriate technologies, cisterns or other water-storage vessels require some initial and ongoing care to work well. Thus one must look at size and siting, construction methods and materials, and ongoing maintenance.

CISTERN SITING AND SIZE

Location is critical. An ideally placed cistern would be buried (underground walls do not need external bracing to withstand water pressure). An ideal water source is a high-quality, gravity-fed, seasonal spring. If additional catchment areas such as roofs are required, sufficient surface area should be available to furnish enough water for ordinary needs; water should flow by gravity and without mechanical assistance if possible. The catchment structure (roofing or other surface) should be made of a material that will not contaminate the water; this is especially important in Appalachian regions where rainwater has a lower pH (acidic conditions) because of emissions from coal-burning power plants. Likewise, the cistern should not be placed near septic or sewage systems or contaminated water. (See figure 29.1.)

In determining what size cistern to build, it is necessary to take into account the size of the household, what other sources

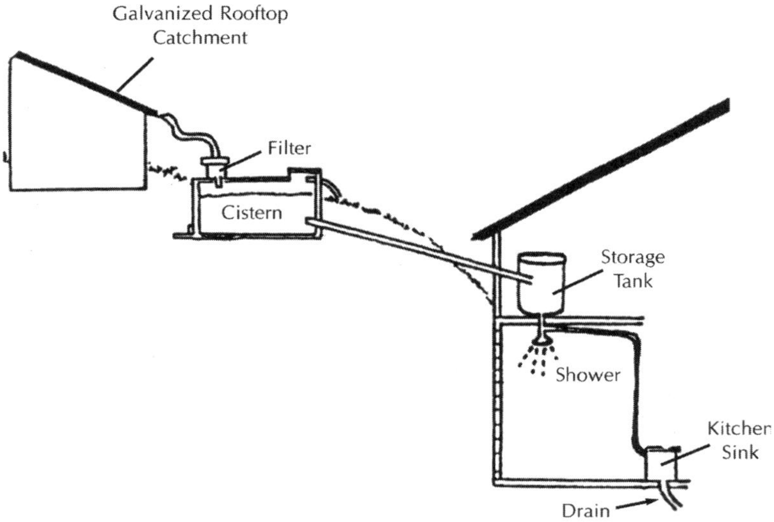

Fig. 29.1. Ideal cistern system

of water will be used, and average rates of rainfall for the area. The cistern next to ASPI's office and gardens at Mount Vernon (see figure 29.2) was not subject to city water regulations curbing garden watering during the time of the 1999 drought; watering continued throughout the dry period from the 8,000-gallon cistern's soft water using conserving irrigation and mulching techniques—and the garden thrived.

If cisterns are to be the only major household water source (nearby uncontaminated ponds, lakes, and creeks can supplement a cistern for bulk needs), rigid water-conservation practices—shorter showers, for example—will be required (see chapter 26). When I grew up in rural Kentucky, two cisterns of less than 18,000-gallon total capacity supplied all our household needs and some flower and garden watering needs as well. The modern national average of 100 gallons a day per person can easily be greatly reduced by household water-conservation measures. A successful cistern user demonstrates to visitors that a high quality of life does not require wasting precious water.

Fig. 29.2. ASPI cistern and sediment tank

Drier regions with long periods without rainfall naturally require larger cisterns. If a household is equipped with composting toilets and some supplementary surface water for bulk uses, a moderate-sized cistern is sufficient. A rule of thumb could be 50 gallons per day per average household of four with a conservative minimum 9,000-gallon cistern capacity. In a region with a 43-inch average annual rainfall, a household could collect a total of 20,000 gallons a year from a 1,000-square-foot catchment with 75 percent water retention. Small cisterns take proportionately larger amounts of time and material to construct, so in general it is better to build larger cisterns to meet all possible needs.

CONSTRUCTION METHODS AND MATERIALS

Generally, for small buildings or summer cabins it is sufficient to buy a commercial 500- to 1,000-gallon preformed, unused septic tank, if easy access to the property is possible. The tank may be plastered—a step we recommend but that is not absolutely necessary—and the interior disinfected; then unneeded holes are plugged. Plastic liners may be used, but often in the past these contained water-soluble solvents and plasticizers, which could eventually contaminate the water. For these reasons swimming pool liners are no longer recommended by governmental agencies as cistern liners.

The simplest cisterns require only commonly available digging and cement tools for the do-it-yourselfer (picks, shovels, trowels, etc.). Some people may be tempted to build an unwalled cistern in nonperc (no leakage) clay soils and plaster the excavated clay walls. While this may be possible in certain dry regions, it is not recommended in parts of Appalachia, because most areas are subject to natural ground shifts, the effects of blasting for construction or mining operations, shrinkage of clay layers due to drought, or natural seepage of contaminated groundwater.

External walls should be of brick or 4-inch concrete solid block, laid in regular fashion, preferably in a round dug-out space. Ferro-cement water tanks can be built at low cost as well. Chemical sealants are not recommended; instead, waterproof with two coats of one-two-three portland cement mortar mix. Plastering should be done from the bottom upward in one continuous application. A plastering trowel is used, and the outer plaster surface is smoothed with a thin cement slurry just before drying to ensure that any fine cracks are sealed.

One of the trickiest aspects of cistern construction is sealing the top or roof to prevent external surface seepage. When cisterns were carved from rock, that medium served as a ceiling. More-creative builders have built arched domes at the top of the cistern, but few today have the skill or the time to do this. Today, cistern builders prefer the preformed containers mentioned earlier. If the cistern is constructed by hand, a wooden collapsible form can be built and 4 or more inches of concrete poured on a standard reinforcing steel (rebar) assembled form. The interior portion of the form is removed piece by piece through a small aperture. The key is nailing the form's floor in such a fashion that all supports and flooring can be easily removed from the interior or exterior after the poured ceiling has hardened.

Two able-bodied builders can complete a 12-foot-diameter, 10-foot-deep cistern using hand digging in normal soils in a few weeks. A 10-foot-deep cistern need only be dug to 6 feet and the excavated soil used as a berm for the remaining exposed 4-foot wall. The greatest need for support is in the lower portions of the cistern. Masonry cisterns should not be exposed aboveground without reinforced steel because of water pressure on the sides.

Not counting labor, a cistern could cost as little as $1,300 in materials (sand, cement, wall materials, rebar for the ceiling, manual water pump, and simple sand and gravel filter). Bear in mind that municipal water systems with pipes, purifying plants, and pumping and storage systems cost much more per household, though governmental subsidies cover much of that cost.

At Point Reyes, off the northern California coast, I observed a lighthouse with an old cistern whose catchment was protected by the bare rocks above the cistern entrance. Such catchment materials are ideal if the initial flow of water is diverted and only the purer succeeding flow is saved. This would be an excellent suggestion for rocky parts of Appalachia, with possible masonry additions.

As mentioned, the growing acid rain problems in our region require additional attention to the selection of catchment materials and the added protection needed for potable water. For uses that don't require potable water (garden and flower watering, laundry, bathing, vehicle washing, and water for pets), virtually any roofing material will suffice, if a normal filtering system to keep out larger particles is attached before water enters the cistern.

For drinking and cooking purposes, extra attention must be given to catchment materials. Even in regions with acidic rain, a shale, tile, or weathered wooden shake roof would be sufficient without added protection. We would not recommend non-enamel-coated aluminum roofing catchment for potable water because of current acid rain conditions. Weathered galvanized steel (without a lead-based painted surface), baked-on enamel aluminum, and popular asphalt shingle roofing may require additional commercial water purification units; bear in mind, too, that older asphalt roofing may contain asbestos. New roofing should be allowed to season or weather for a period before being used as catchment for potable water.

A standard cistern filter is a box that contains washed sand and gravel layered in such a fashion that larger particles carried from the catchment are filtered out before entering the cistern. (See figure 29.3.) In a large cistern attached to a seasonal spring, ASPI has built a smaller buffer container so that any debris will be caught in that unit before the water enters the cistern (a pre-filter system). We suggest that in areas affected by giardia and other parasitic or infectious disease organisms the potable-

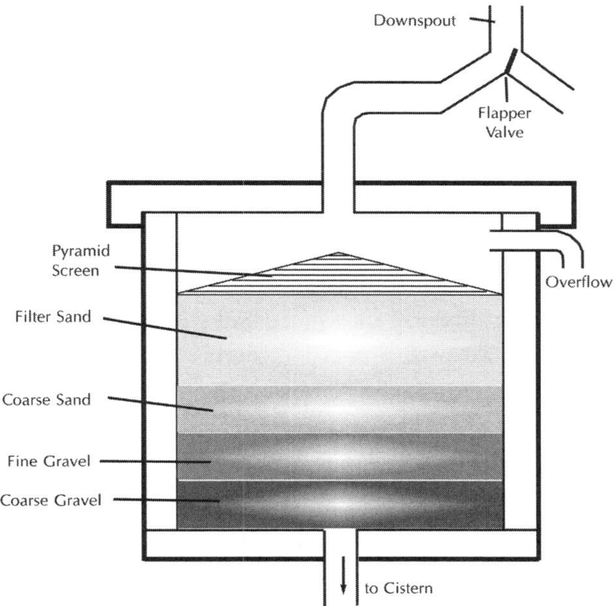

Fig. 29.3. Rainwater filter

water faucet be equipped with a final commercial water filtering system.

CISTERN MAINTENANCE

Water from a properly maintained cistern poses no more risks than water from virtually any other source. All drinking water must be protected, whether from municipal or individual sources. Individualized sources such as cisterns are denigrated by some regulators because the sources are not centralized and extra effort is required to visit each site. Tightly sealed cisterns in areas where the water is not contaminated can be quite safe and do not need chlorination, although using commercial water purification units may be a safer practice.

Experienced cistern owners offer these suggestions for maintenance:

> Catchment and gutters must be kept free of leaves (a standard house maintenance practice with or without a cistern). Cutoff valves should be installed in the downspouting and left always in the cutoff position, except when the rain is sufficient to have washed any particulates off the catchment surface. That first cut of water can be diverted and saved in a special tank for watering the garden in dry weather. My Kentucky home place had a cistern with such a system and accumulated less than ½ inch of sediment in over half a century of continuous use.
>
> The border of the access-hole cover should have a lip slightly higher than the concrete ceiling of the cistern to prevent surface seepage. The concrete cover should be lightly sealed with caulk or cement that can be easily broken and removed if it is necessary to enter the cistern proper for general repair or in case of unexpected problems.
>
> Particle-removal filters should be replenished with sand and gravel every several years, or more often if the structure with the catchment has a masonry chimney used for space heating. Water should not be allowed into the cistern when such chimneys are in use, or until after a prolonged washing by a heavy rain after the heating season.
>
> After earthquakes or major land disturbances such as rock slides or blasting in the vicinity, the cistern should be checked for cracks.
>
> Keep mechanical cistern pumps oiled and replace parts when needed.

POPULARITY OF CISTERNS

Cisterns have taken a bad rap in America during the past half century as the emphasis on improving water quality has grown. Some of this dislike is justified because builders of older cisterns did not give sufficient attention to sealing off the tops of the containment vessels, allowing contamination from ground seepage. Because of cisterns' poor reputation, bureaucratic bias may cause regulators to define all cisterns as bad, before even investigating them—especially because of the extra effort required to visit each site.

Cisterns will not become popular until they receive more authoritative endorsement and sponsorship along with some public or private research into better catchment materials and purification methods for potable water supplies. Very few regulators object to using cistern water for purposes other than human consumption. All water systems have risks attached, even monitored municipal water systems. A 2002 study by the U.S. Geological Survey found that a soup of drugs, pesticides, and residues of personal products (antibacterial soap, estrogen replacement drugs, birth control compounds, etc.) is slipping through filtering processes at municipal water treatment facilities. Today, people are becoming more aware of municipal drinking water problems. Male fish living near the London, England, wastewater treatment plant are producing eggs due to estrogens from birth control pills. Chemicals forming in a contaminated chlorinated medium are adding to human health problems. It is not surprising that the sales of commercial filtered spring water have skyrocketed—although they are not necessarily of better quality than water from other sources.

Well-constructed cisterns can furnish soft, clean, good-tasting water if the proper steps are taken. This message is needed in many parts of the world, especially in places like Bangladesh, where as many as 35 million inhabitants face arsenic poisoning from shallow ground well water used for drinking and cooking.

The problem could extend to other Asian countries like India and China, where arsenic contamination in groundwater could affect 100 million people. "Poisoned Waters," an insightful article in *World Watch* (January/February 2003), focuses entirely on well drilling and related problems and testing and simply dismisses as too expensive rainwater containment in a land averaging 80 inches a year. Really! It is becoming increasingly imperative that environmental consciousness be coupled with appropriate technology methods and solutions. Cisterns could be a partial solution to China's and Bangladesh's water problem, if there were only a will on the part of policymakers to look in that direction. That is also true for Appalachia.

CHAPTER 30

Irrigation and Water Conservation

We hear the sound of laughing waters from our home in the mountains year around. That sound is pure joy and the water is cool and refreshing. Fresh mountain spring water is such a gift and blessing to our lives.

When the power goes off, we continue to have free flowing water to drink, wash, and bathe, thanks to our gravity-fed spring water system. We use it to water our solar greenhouses, gardens, herbs, flowers, and berries. In dry weather the sound of running water becomes more gentle and quiet, but the stream never stops flowing.

Fresh, pure water is a limited resource and we feel compelled to use it responsibly. We use a passive solar composting toilet to conserve precious water resources, consuming less than one-half the amount of water used by a typical family. We wash our dishes by hand—no dishwasher needed, thus conserving water and energy. In our home we know that conservation extends to all of creation and we try to be a living example of conservation through our everyday water use and practices.

Pat Gallimore, Long Branch
Environmental Education Center

In normal years central Appalachia is blessed with about 47 or more inches of rainfall, ample for most domestic and commercial needs in the region. However, the people of Appalachia descend from pioneers who labored to carry buckets of precious drinking and washing water from a spring or hand pumped their water from a cistern. The pioneers' offspring may easily forget water conservation, however, when a slight twisting of the tap can initiate a plentiful supply of municipal water. If we are committed to being Earth-friendly, we must act as though conserving water is good discipline; water conservation shows respect for resources; saves energy required to pump, store, deliver, and restore water; and reduces the quantity of wastewater. It also makes higher quality water available for other purposes.

Often the home owner during severe drought is reduced to a triage system of watering taken from battlefield medical practice, where the wounded were divided into those critical but recoverable, the maybes, and those who were dying and would not respond to emergency treatment. Garden triage means watering the most sensitive plants, such as recently planted seedlings or those like melons that need water most urgently, then tending to longer-lived plants that can survive on less water, and leaving out the rugged or already-bearing types (okra, onions, mint, and Jerusalem artichokes).

SIMPLE IRRIGATION

The more water for crops that is available, the better the chance that gardens will yield good, wholesome produce. Late summers and autumns are often drier times, but experience has taught us to be prepared to water crops throughout the growing season.

The elaborate water delivery systems of the water-limited Great Plains and Far West are normally not found in Appalachia. Here the problem for gardeners is to get water to points needed from available sources—cisterns, ponds, streams, creeks, groundwater (wells and springs), and municipal water systems. The delivery system is often a good bucket and a strong back to keep the recently planted fruit tree alive or the flowers fresh. The speed at which a vase of wildflowers sucks up water is a lesson to us in how much it takes in nature to keep plants from wilting and dying during summer's heat. Gardeners should have sufficient sources of water, a handy delivery system, and application methods that minimize the quantity of water .

A single dry spell will convince many that native plants usually endure a drought far better than most exotic varieties. During dry times, lawns of exotic grasses and planted flowers are some of the first to wilt. In most places, when the water supply gets low, the first thing mandated by local governments is to stop the watering of lawns. A better approach is to select plants in the initial stages of landscape design that are native and require less additional watering in the initial stages of landscape design. Water needs are reduced further by mulching both native and exotic plants. Placing a layer of the mulch around the base of the plants suppresses weed competition, stimulates earthworm activity, and can help to reduce the loss of moisture through evaporation.

Watering certain areas by hand is the main practice of choice in Appalachia, provided the number of plants is not large and the water is plentiful (see figure 30.1). A gardener living near a creek could perhaps dip enough buckets to water a relatively large bed, but that could get tiresome during an extended drought. Gardeners with limited water soon learn that pure water is best but can be replaced by gray water—bath- or wash water—or by diluted urine, which is rich in nitrogen. Timing becomes important, so watering should not be done in the heat of day but should be relegated to early mornings or, better yet, to the late evening for maximum absorption before the hot sun of another day dries out the earth. Watering with rainwater collected from the roof in a tank or cistern is an exercise in appropriate technology. However, the abstemious plant caregiver pours water at the base of each plant that needs water and does not wet the entire plant through aerial spraying. A water-intake pipe in the ground or a drip bottle for tomato, pepper, and cucumber plants is helpful for minimizing water waste.

Having said that water should be poured at the base of the plant, it's important to remember that we are talking about small-

Fig. 30.1. Pat Brunner of Berea, Kentucky, waters a raised garden bed.

er gardens, which can be handled by individuals seeking daily exercise and a renewed connection with nature. Large gardens, nurseries, or specialty crops may require less labor-intensive watering procedures. Often commercial growers resort to sprinkler systems but have undertaken practices to limit this relatively high-volume water use. They water certain crops more than others, watch the time of delivery carefully, use mulch around plants, and refrain from watering during the hotter parts of the day. Sprinklers can deliver needed water over a shorter period of time and can use farm pond and other lower-grade water with greater ease than drip irrigation systems, which conserve far more water. Needless to say, some of the water misses the target plants through evaporation or unnecessary watering of paths, field edges, or spaces between plants.

DRIP IRRIGATION SYSTEMS

Modern drip methods have roots among Native Americans that go back thousands of years. The Pueblo Indians in the Southwest practiced "pitcher irrigation." Near the plants to be watered, large unglazed earthenware jugs were buried with mouths at soil level. Water in the covered, filled jug seeped slowly through the porous pottery to the plants. This method can be used today. Gardeners can also use less-expensive 1-gallon plastic containers of water with small holes in the corners to allow water to drain out and water plants in the direction of all four corners over a period of time, depending on the aperture size. However, these containers must be periodically refilled by hand.

The Symcha Blass drip or trickle irrigation technique, especially when drip irrigation plastic tape is used, has many advantages. It requires 30 to 50 percent less water and smaller pipes than sprinkling methods because of lower pressure, and little or no energy. This method provides better control of soil moisture and more precise moisture placement; avoids wetting plant foliage and fruit, which may encourage fungus; allows paths to stay dry and thus limits weeds; is less labor intensive; allows harvesting in fields while irrigation continues; is not affected by wind as is field sprinkling; and allows for application of organic fertilizer through the water system.

The major disadvantage of this largely plastic irrigation system is that it requires quite clean, sediment-free water, lest the small apertures in the emitters clog with soil or organic particles. Those using this system generally turn to municipal water for purity, though it is often chlorinated, which can be detrimental to the growth of various garden plants. Other problems include soil moisture being limited by volume of soil wetted and orifice discharge. The distributing tapes can be subject to rodent, insect, and mechanical damage. The entire system requires a higher level of design and management than other systems. Finally, the initial and annual investment is sizable , since many polyethylene tapes are discarded after one year and accumulate in waste heaps. With care these T-tapes can be used for the local delivery system a second year, but they are often damaged by the stresses of being in the ground, clogging, and exposure to weather and sun.

Large drip systems involve a sizable investment depending on the amount of ground that must be irrigated. For any system, besides the water source, there must be a straight or self-priming centrifugal water pump and power unit (gravity systems are generally not feasible), a primary filter at the pump and secondary filters at the distribution head, pressure regulators and gauges, control valves, and main lines of heavy-duty piping that can be placed underground. In larger operations, distributive piping (ordinary garden hose or flexible ½-inch plastic tubing) requires laterals of ½-inch or ¾-inch polyethylene pipe. Tape emitters or local delivery systems for trickle or drip end use should be low pressure but capable of counteracting the pressure differences caused by topography and friction loss. Spacing and aperture size are selected depending on crop needs, topography, and pressure of the system. There is flexibility in laying out these emitters on mountainous or terraced terrain as well as on flat surfaces (see figure 30.2).

Point source emitters, used to water trees, vines, or ornamental crops, are more rugged and can be reused year after year. They can be attached to a pipe or installed in line. Some of these systems have discharge rates from ½ gallon to 10 gallons per hour; delivery amounts will depend on length of time of water flow.

WATER CONSERVATION

Irrigation is the major water use in America, with agriculture and related industries consuming more water than any other sector. However, the demand for potable household water is growing in all parts of the nation, mainly because people are using more and more domestic water. Whether water is relative-

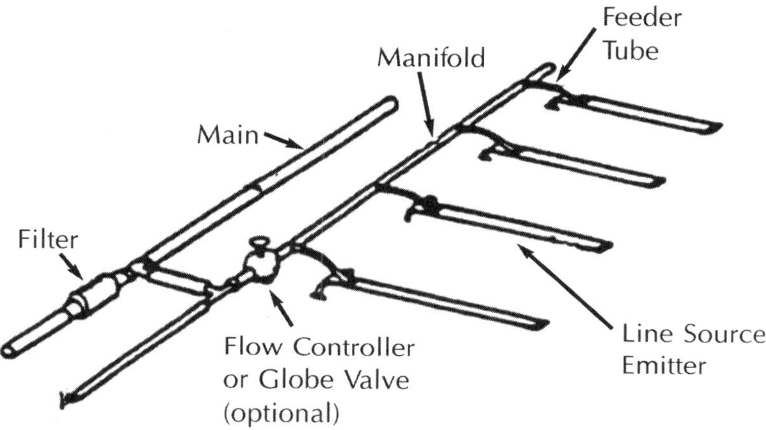

Fig. 30.2. Typical drip irrigation setup

ly plentiful or in short supply, domestic users need to conserve because ecological awareness and conservation of resources go hand in hand. We must learn that small amounts of water can still do the same quality job as larger amounts. Water takes energy resources to gather, store, purify, and deliver, and saving water saves money. Following are a number of water-saving practices.

Install a composting toilet. No device will save more water, since it needs none for flushing. There are estimates that substituting composting toilets for flush toilets would save half of all water used in the home. (See chapter 27 for details.)

Use low-flow devices. This can save sizable amounts whether installed in the kitchen or utility faucet or in showers and flush toilets. Hand washing does not require large amounts of water; small amounts used over a longer period of time can do the job equally well. Virtually any supermarket hardware section has faucet aerators that create a more dispersed spray, which adequately removes soap residue. Governmental regulations are making such conservation devices mandatory in new home construction. Similar water-saving devices are available for car washing and other cleaning needs.

Consider recirculating water fountains. Ornamental fountains and waterfalls are often found in homes and public and private buildings. However, the bubbling, gushing, falling, and cascading of water can result in sizable water losses through leaky drains and evaporation. Much of this is saved through recirculating the water and shading the areas where the water activity occurs with trees, vines, and trellises. Light-colored surfaces around such devices also help reflect the sun's rays and reduce evaporative loss. Artificial shading may also enhance the sound of the water in more enclosed spaces. Water circulation devices can be put on a timer or turned off when traffic is light or after business hours. Solar water pumps are especially useful, for they operate in sunlight, when the circulating water is more deeply appreciated (see chapter 1).

Use appropriate personal washing practices. In former days people bathed less frequently and in relatively small amounts of water. In my family we washed our faces and hands before each meal, but the major weekly event was the Saturday night bath that each of us took in the same large galvanized tub next to the kitchen coal stove. For each of our baths we got a half gallon of warm water from the kettle and warm water reservoir attached to the coal stove, poured over our backs in showerlike fashion. We washed with the several gallons of soapy water left over from the previous washings. We were scrubbed quite clean with lye soap, yet a half dozen people required about the amount of water used by one modern shower. Though this method is out of fashion today, teeth can be brushed by wetting and rinsing the brush rather than running the tap full blast. We can shower by wetting down, soaping down, and rinsing off in a water-conserving fashion, using about 3 and not 15 gallons, as does the average modern shower. We can follow good hygienic practices with far less water.

Conserve lawn water. Watering lawns can be a very wasteful practice and is generally the first to be stopped in a time of water shortage. A ground cover requiring less water can be substituted for grass. Edible landscaping or wildscaping can also be planted, or lawn areas can be allowed to return to woodland. Another domestic water-conserving practice is to use a broom to clean sidewalks, decks, and porches rather than a strong jet of water from a hose.

Use dehumidifier water. The eastern half of the United States is relatively humid, and dehumidifiers are frequently used in homes. These devices as well as air conditioners can condense gallons of water each day, and the condensate can be used to water plants or for other nonpotable purposes.

REFLECTIONS ON WATER USE

Ultimately, water availability and quality affect all of us. How and when should one irrigate, when 90 percent of all U.S. water use is for agriculture? There is also the issue of the use of agricultural or lawn pesticides and heavy doses of fertilizers that contaminate the runoff for people downstream. Even in fairly nonagricultural Appalachia, water removal from a free-flowing stream could affect the quality of aquatic life and thus reduce the quality of the fishery and the recreation potential of a given watershed. It is unconscionable to use large quantities of pure municipal water for irrigation—and yet this is what drip irrigation tempts gardeners to do. And drip irrigation advocates accept that municipal water is the best water source and running streams the least desirable because of sediment loading. If drip irrigation methods were used for moderate- or large-scale agricultural irrigation, these systems would rapidly use up the entire municipal water supply. Thus one understands the continued popularity of sprinkler systems, which can use lower-quality untreated water.

Water is what makes Appalachia and all of eastern America green. People coming from the dry West always feel refreshed when coming into the greenery of the eastern mountains. Much of the coolness, beauty, and freshness of this landscape is due to the blessings of flowing streams and gentle rainfall. Of course there can be too much of a good thing—frequent and disastrous floods are particularly likely to occur in Appalachia where the protective vegetative cover has been removed. The effects of surface mining and forest overharvesting are part of the flooding problem. A sensible water conservation measure is to keep forested lands healthy so that the rainfall will be absorbed slowly and deeply. Ultimately, good land conservation is good water conservation. Water quality and quantity are always community concerns and require the full participation of all who are committed to saving the water commons.

Conclusion: An Appalachian Appropriate Technology

The past decades have seen considerable progress in Appalachia and elsewhere—in the field of appropriate technology. From recycling of wastes to energy conservation in "green" building design, we are getting much better at meshing environmental constraints and economic needs with technologies for meeting human needs.

Appalachia has an important role to play in the future of appropriate technology because it is a very special land with a long tradition of understanding the complex relationship between meeting human needs and exploiting the ecosphere's limited resources. Appalachia and appropriate technology are natural historical partners. This partnership must be expanded in the years ahead to ensure increased economic viability and decreased pollution of this strikingly beautiful part of the world.

Arthur Purcell, director,
Resource Policy Institute, Los Angeles

This final chapter does not attempt to predict the future of Appalachia in any way, for that is impossible. Rather than try to predict the future, let's try to influence it for the better. Our region needs to improve in quality of life, environmental integrity, and economic health. Appalachia does not stand in isolation but is part of a nation and world seeking environmental and economic integrity. Concerned Appalachians are challenged to make this region of relatively limited resources into a model of what leadership in this affluent nation can do to use resources wisely. People can provide for their own basic needs without sacrificing the environment. The region can answer some terrorist threats by offering decentralized energy sources, which are far less vulnerable targets than large-scale power plants and electrical grid systems. Finally, Appalachia can take the lead in showing that higher quality of life can be attained by all and thus defuse the volatility of systems where lower-income people see little promise in the future.

Appalachian people have traditionally been hospitable; they give time to others and truly want to know how people are; they open the door to full community participation. Many Appalachians have seen some of the traditional ways threatened by the desire to be entirely immersed in American consumer culture. Does the region have to follow the crowd, or can 9/11 and its aftermath be a wake-up call to reexamine the wasteful lifestyle that troubles many conservation-conscious Americans?

The pressure to safeguard petroleum sources in the volatile Middle East—needed to support what President Bush calls our "oil addiction"—is part of what drives our involvement in that region. An alternative, Earth-friendly approach to meeting energy needs would be far less costly in time, resources, and personnel. A soft-technology alternative approach would reduce dependence on oil and other nonrenewable fuels as well as the military security needed to ensure the flow of overseas petroleum resources. This renewable energy approach, if taken seriously, requires models of self-reliance—a characteristic embedded in Appalachian culture for decades. Logically, this region should have a leadership role in lowering the need for costly imported fuel and other resources. Self-reliance costs less and yields more and instills confidence and self-respect, and Appalachia can lead the way.

IMPLEMENTATION

The thirty examples of appropriate technology discussed in this book have enjoyed some approval and success, at least on the individual level. But the critical question is, Will the examples work on a regional level? Will they lead people to follow the examples and incorporate them on a local community level?

We were part of the heady "appropriate technology" era of the late 1970s, being major contributors to the Southern Unity Network for Renewable Energy Projects, a coalition of organizations working for an alternative energy and friendly technological approach for the South. Members of the network met monthly and coordinated the use of available federal resources to bring about grassroots awareness of changes sought for a better quality of life. At that time, we would be quick to say, "Yes, we can put these examples to work right now." But only a few years later, in 1981, these federal appropriate technology programs were abruptly shut down and have never been restarted during either Democratic or Republican administrations. The nonprofit organizations we worked with either withered away or retrenched for the long haul. We the authors sought national, regional, and local environmental programs that would extend self-reliance and sustainability to the nation as a whole and would ensure our organizations' survival. Thus emerged our Environmental Resource Assessment Program (see www.earthhealing.info), and we have seen recommended approaches come alive among assessed groups.

Over time we have come to take a long-term, realistic view and say that the future success of the technologies presented here depends on more than individual efforts. Some eight of the thirty are primarily individual; an equal number require local community coordination; another eight areas are more focused on regional activities; and six areas require some national methods and programs. We group the thirty subject areas into these four categories and highlight one source of assistance for implementing each practice, but we recognize that all have multiple areas of influence. For instance, water conservation is a community concern but involves individual habits, regional water allotments, and national water-allocation policies in resource-short sections of the country.

Individual Level

The following technologies are appropriately implemented primarily at the individual level:

Edible landscaping (chapter 8). Models of good landscaping with fruit and nut trees, herbs, and vegetables can now be found around the country and can easily take root in our region as well. The literature is there for the taking. *Organic Gardening, BackHome, Mother Earth News,* and other national periodicals have furnished numerous examples of what can be, and is being, done. Features in local newspapers and regional news outlets can help popularize the practice.

Solar greenhouses and extenders (chapter 11). Greenhouses were built in the 1970s through government-funded local workshops, which proved quite successful. The payback included acquired skills and knowledge for individual participants, produce from the greenhouse, and additional solar heating for the attached residence or business. More and more people are seeing that the addition of an attached greenhouse can enhance the property's value as well, even in the absence of government programs.

Retreat cabin sites (chapter 18). Quite often people have scenic property and wish to preserve the view; they can be persuaded to build lower-priced, primitive buildings on the land that could be rented out to those wishing a mountain experience. The best way to select a location would be through an environmental resource assessment. Our Web site gives a list of places where such facilities are now functioning and can be visited.

Cordwood structures (chapter 22). A cordwood building is so low priced that virtually anyone who has property can afford to build one according to basic plans available from the literature listed in the Resources section. For those without building skills, the best advice is to recruit a local building expert to assist with plans. Tapping the local pool of farmers and skilled builders is often overlooked in this region. It may be possible to recruit volunteer labor for the construction, as I did in 1982.

Yurts (chapter 23). A yurt is a tricky little building to construct even with plans from the Yurt Foundation. It does not take long, but on the day the basic structure is raised, a number of helping hands would be most welcome, along the lines of the barn-raisings in nineteenth-century America.

Composting and vermicomposting (chapter 25). Composting has always been done on some level in this region. As elsewhere, special attention must be given to dramatically reducing the domestic garbage volume and returning organic matter to the land. Youth may lead the way, for a number of regional school programs are starting to emphasize domestic composting through student environmental education projects.

Composting toilets (chapter 26). Individuals must choose to install such devices, but the requirements of local and state agencies are also very important. Several of the Appalachian states are more lenient than those outside the region. A paramount requirement for popularizing these devices is small-scale demonstration of their success—and this is tricky, because some undersized composting toilets in public park facilities have not been able to cope with visitor use. Demonstrations are better done at an individual or small-group level where oversight and management will be ensured.

Cisterns and water catchments (chapter 29). An individual decision to build or renovate a cistern will prove helpful for conserving water, especially in drought conditions. Private and government plans and technical instructions will be useful in the construction phase. Use of local expertise is again important. Information on ways to avoid the pitfalls of older cisterns (which allowed seepage of surface water into cisterns) is essential as well.

Local Community Level

Local communities are the best place to implement and promote these technologies:

Solar photovoltaics (chapter 1). During 2005 the Long Branch Environmental Education Center held highly successful hands-on workshops for individuals who desired to build their own portable solar generators. This mobile device can collect solar energy, store energy in batteries, convert direct current into 120 volt alternating current through an inverter, and deliver solar energy to construction sites, residences, or work areas. These devices can be used where community residents have lost grid electricity due to power outages or disasters. The people attending the workshops learned basic solar principles and discovered that they can own their own solar energy source at relatively low cost (less than $2,000 for a solar generator).

Food preservation (chapter 7). Food preservation may be regarded as primarily individual, and it certainly involves personal choices, but it can also be done in community, following the example of pioneers. Community canning during the summers was popular at school kitchens during World War II. The Appalachian Sustainable Agricultural Project in North Carolina and a similar project in Tennessee are promoting a community approach to food preservation. While not yet fully extended throughout the region, this traditional approach to food preservation, which has lapsed into disuse, offers immense promise.

Intensive and organic gardening (chapter 9). The indigenous Cherokees and later-arriving Appalachians excelled in good local intensive and chemical-free gardens for decades. The growing number of farmers' markets can help boost the popularity of these gardening methods by offering an outlet for surplus crops. These budding markets also offer opportunities for educational exchanges among like-minded people. The federal government has a small marketing program as part of its agricultural budget.

Wildcrafting (chapter 15). Gathering wild plants, roots, seeds, berries, fruit, and craft-related materials is a strong tradition within the Appalachian region, but often the outlets for products are not well known. Popularization of products is a key to success. County fairs and festivals offer ideal outlets for the wildcrafter's products. The woodcrafter's festival in Washington, Kentucky, credited to the inspiration of my father, a local wood-carver in his retirement years, is an example of the development of economic outlets for a particular specialty.

Natural cooling (chapter 20). The best way to reduce air-conditioning bills is by planting (or not cutting down) shade trees and replacing existing dark roofing with lighter-colored roofing. Local communities could plant trees as part of National Arbor Day celebrations. We have been encouraging church youth groups and other community service groups to plant trees in honor of deceased members. This can get people into the habit of planting trees and encourage them to plant trees as part of annual celebrations and homecoming events.

Native building materials (chapter 21). Buildings should be constructed using materials at hand—a traditional practice throughout human history. This is a more creative and far less expensive approach than patronizing large chain building supply places, which draw materials from many distant places. The emphasis is thus on buying from local commercial establishments (e.g., sawmills, lumber companies, quarries) or retrieving the materials on one's own from local sources. Community efforts are critical here in sponsoring, financing, and selling local products in a cooperative fashion.

Recycled, salvaged, and deconstructed materials (chapter 27). Virtually everyone recycles some things at the individual level. But our own reuse of materials often involves community participation, for accumulated salvaged or deconstructed materials require a gathering place at the local community level. In less densely populated counties it is difficult to maintain facilities for low volumes of used construction materials. However, existing or planned recycling center programs are ideal for such gathering and exchange of materials.

Irrigation and water conservation (chapter 30). Water quality and quantity is a community concern as well as an individual issue. Individual households can use water-saving techniques suggested in the literature. Often, however, especially in times of drought, the community must dictate how much water is used per household and when it can be used. Community education programs through newspapers, radio, local Internet networks, and television have a great part to play under such circumstances in conserving water and maintaining the quality of local sources of drinking water.

Regional Level

The followÏing technologies are best implemented at the regional level:

Wind power (chapter 3). One may speak of wind power on an individual or a commercial utility basis, but these need not be totally separated. Through net metering, wind sources can come from the small-scale units that are now being implemented. Furthermore, wind farms are now becoming popular through the Tennessee Valley Authority (TVA) and others. In the past, soft technologists worked hand in hand with TVA, especially through exhibits at the Knoxville World's Fair in 1982. If these partnerships are revived, regional agencies and utilities could encourage the installation of individual systems to feed surplus energy into regional grids.

Solar heating applications (chapter 5). Renewable energy expositions are now appearing in various regions (e.g., ASPI's Bluegrass Energy Exposition). These enable participants to meet people who are knowledgeable about a wide range of solar and other renewable energy applications, to observe and talk about products firsthand, to speak with commercial agents, and to obtain pertinent literature. The annual Southern Energy and Environment Exposition (www.seeexpo.com) in Asheville, North Carolina, has attracted thousands of people from the southern Appalachian region. This exposition has helped make solar energy popular, so that is no longer perceived as pie in the sky.

Regional heritage plants (chapter 10). Efforts at both national and regional centers to preserve plants have been highly rewarding. Exchange of heritage plants now occurs at local marketing and herb growers' groups and at such Appalachian organizations as Rural Action in southeastern Ohio. Long Branch's heritage apple orchard and grafting workshops fit into this concern to preserve some of the important heirloom varieties and diversity in our food crops. More and more people want access to heritage plants from this region.

Wildlife habitat (chapter 12). The designation of areas as wildlife reserves, such as bird sanctuaries, can be done at various levels. More areas will be so designated through coordination by state wildlife agencies. Long Branch's 1,600-plus acres are designated as a "protected natural area and wildlife reserve." The driving force in the region is the development of Appalachia and the fragmenting and closing off of larger stretches of forested area, thus limiting wildlife migratory routes and habitats. This tendency must be countered.

Nontimber forest products (chapter 13). Each region has a variety of materials that could be harvested at sustainable rates and could generate some income for property holders. Each state land grant university has programs to which property owners could turn for ideas. NTFP regional seminars, especially in western Virginia and western North Carolina, have been quite popular and beneficial. Extending this format to all parts of the region is highly desirable.

Silvicultural practices (chapter 14). With enough ecological literacy, individuals can engage in good silvicultural practices on privately held forests. However, state guidelines for good growing and harvesting procedures can go a long way to preserving and improving the quality of Appalachian woodland. While private groups like the Forest School (see chapter 14) can teach such practices, state forest agencies must work to extend the practices by promoting tree-planting operations, furnishing seedlings, popularizing good forest management practices, holding field events, monitoring forest health, launching campaigns to stop regionwide tree blights and pest infestations, and enforcing forest protection regulations.

Constructed or artificial wetlands (chapter 16). In western Kentucky, artificial wetlands are quite popular and thousands are in operation. Their popularity has not yet extended to Appalachia, but spreading the practice could be the task of the Cooperative Extension Service and water quality agencies at the regional and state level in all the Appalachian states. It is possible to observe these wetlands at rest stops along Interstate 40 in Tennessee, but greater publicity could help popularize them still further. Appalachia's problems with "straight pipes" (feeding wastewater directly into streams) could be countered in part by creative ways of building artificial wetlands on limited space and in elevated slopes as was done at the ASPI solar house. Such wetlands can be beautiful and fully integrated into the landscape.

Transportation (chapter 24). Individuals can choose efficient and affordable vehicles; a conscientious local mechanic can give dependable individual service; communities can furnish public transportation for senior citizens. But overall improvement in transportation will depend on larger areas of cooperative endeavor. The regional planning sessions held at Elcorn City, Kentucky, and Breaks, Virginia, in April 2005 focused heavily on transportation modes (roadways, parking, walking trails, biking, and public rail facilities). Similar regionally sponsored events are essential to getting Appalachians thinking about and moving toward alternative modes of transportation. At the same time, the federal government must put teeth into nationally mandated auto efficiency standards and establish more incentives for hybrid vehicles, solar electric vehicles, and hydrogen fuel cell service centers for the next generation of private vehicles.

National Level

Some technologies could benefit from assistance at the national level. These include:

Microhydropower (chapter 2). This renewable energy source does require a certain amount of knowledge about materials, water sources, conduits for water, and electrical hookups. Expertise in Appalachia is present along with abundant water resources. Information is readily available on the Internet to assist individuals and communities in overcoming barriers to tapping free-flowing streams without doing damage to the environment. Federal information programs and tax incentives could augment these individual and community efforts.

Wood heating (chapter 4). A national strategy for wood heating? Yes, because the pollution due to wood smoke can be greatly reduced with current EPA efficiency standards. Older, inefficient fireplaces and stoves have hindered the popularization of this long-existing soft technology. Today, many areas of the country are sensitive to local air pollution problems and thus require highly efficient wood heaters. The technology is in place and the products are readily available; more attention must be given to encouraging their purchase through publicity and even tax incentives.

Shade trees and windbreaks (chapter 6). Federal programs establishing windbreaks and shelterbelts were strongest during and right after the Great Depression and especially in the Dust Bowl era on the Great Plains. It is quite possible that modern windbreak legislation could become part of an amended National Energy Plan, for more than agricultural reasons. Trees provide summer shade and protection from cold winds for homes, yet many people do not see that tree planting could be an integral part of a national energy conservation strategy. Tree placement and care in public areas could be the responsibility of a public-private partnership in local communities.

Land Reclamation (chapter 17). Land reclamation in Appalachia may require no new legislation but enforcement of existing laws. This enforcement could be better achieved through citizen pressure and monitoring to ensure that local reclamation sites are properly planted only with native non-invasive species according to the regulations under the amended U.S. Surface Mining Reclamation Act of 1977. Preservationists fighting against invasive species and regional environmental groups could join forces to ensure good land reclamation throughout the region.

Energy-efficient passive solar design (chapter 19). The American Solar Energy Society, the Interstate Renewable Energy Council, and a host of regional and state organizations sponsor annual tours of solar homes in October, since every "green" home should be a demonstration site. In recent years, visitors have been able to come and see for themselves how others live or work in well-designed solar-energy environments. This firsthand approach has popularized the resource in all parts of our country.

Ponds and aquaculture (chapter 28). In the twentieth century the U.S. Department of Agriculture was highly effective in furnishing expertise and financial support for farm ponds in all parts of the United States. That farm pond program needs to be revitalized to help design and build more water impoundments that could be used for irrigation, water for livestock, fee fishing, flood control, fire safety, and aquaculture on family farms and in small communities throughout Appalachia.

TWO SCENARIOS

Let's consider two scenarios, one the continuation of business as usual with a heavy dose of modern technology and one with applications of Earth-friendly technologies. And in between, we offer a hopeful example.

Business as Usual

Current environmentally degrading practices show little respect for either land or people, accelerating the pace of deterioration. Among the anticipated effects are the following:

> Electric energy bills continue to rise, as does demand for heating and cooling. Increased numbers of coal-burning power plants, even with state-of-the-art pollution controls, continue to contribute to the haze and air pollution.
>
> Food bills are a major drain on household budgets, for only a small portion of food is homegrown. Locally grown fresh produce remains hard to find in smaller Appalachian communities. Most of the produce is imported despite growing concern about quality and chemical contamination due to lax surveillance of imported foods. And these imported fruits and vegetables are beyond the budgets of many in Appalachia. A heavy dependence on relatively lower-priced fast food results in high rates of obesity and obesity-related illnesses among both young and old.
>
> Poor forest harvesting practices continue to plague the mountain regions. Rapid runoff of storm waters exacerbates damage in flood-prone regions because slopes are no longer protected by forest cover. Nontimber forest products continue to be haphazardly harvested, to the detriment of the health and propagation of wild ginseng and other species.
>
> Land is being developed at a rapid rate; land degraded by strip mining is poorly reclaimed. These practices degrade the

appearance of the area and harm the viewscape in many places, to the disadvantage of tourism.

A growing number of new residences are manufactured and mobile homes, which depreciate more rapidly than on-site permanent construction. These prefabricated buildings are more susceptible to wind and storm damage and have many problems with leakage and indoor pollution due to poor ventilation and off-gassing of synthetic materials.

Narrow, crowded secondary roads exacerbate school scheduling difficulties in winter, contribute to many traffic injuries and fatalities, and suffer damage by overloaded coal trucks. Roadside walking and biking continue to be quite dangerous. Off-road vehicles continue to take a toll in both driver and passenger injuries and property damage, continuing the trend from the early years of the century, when Kentucky and West Virginia had the country's highest injury and death tolls from these vehicles. The only bright spot is a reduction in fuel consumption resulting from rapidly increasing gasoline and diesel costs.

Wastes continue to accumulate on roadsides and litter remains a problem, even with such aggressive programs as the Pride Program in Kentucky. The severity is intensified by the truckloads of used clothes and furnishings discarded from affluent homes in other parts of the country and sent to Appalachia, where they either wear out and are discarded or remain unused by local people. Waste-disposal costs for these items continue to mount.

Water quality deteriorates throughout the region, and residents purchase more potable water from private corporations at about $2 for every 3 gallons (2006 prices). Because of a lack of secondary, nonchlorinated water supplies, municipal water of barely potable quality is used for flushing toilets, doing laundry, and watering lawns and garden plants.

A Recycled Town

Is it possible to avoid these outcomes? One example can show what can happen when community members work together to rebuild their area and boost civic pride, in spite of a less-than-perfect history.

I grew up a mile from Washington, Kentucky, the third-oldest incorporated town in the commonwealth and the oldest Washington in America (it is now part of historic Maysville). Washington was the site of the state's first municipal water system, the first post office serving the Northwest Territory (1789), and the first bank "west of the Alleghenies." It was the birthplace of Confederate general Albert Sidney Johnston (b. 1810) of Shiloh fame and also the site of the Marshall Key House (1807), where Harriet Beecher Stowe was staying when she was taken to a slave auction, an event that helped inspire her to write *Uncle Tom's Cabin*. Washington is also the site of Federal Hill, the home built by Thomas Marshall, brother of Chief Justice John Marshall.

And Washington has its imperfect history. The town was the home of prominent citizens who went over the Ohio River in a posse to recover slaves who escaped from Kentucky and others caught in the raids across the river. Washingtonians were armed with weapons and backed by federal laws that permitted their ventures. Captured African Americans were then taken back across the river to Washington and jailed. These acts were featured in Ann Hagedorn's book *Beyond the River: The Untold Story of the Heroes of the Underground Railroad*. The book's contents came as a shock to me, for I had always been proud of my hometown and generally passed over its relationship to the "peculiar institution" of slavery. The jail has been long gone, but one local resident said she possesses the lock and key, which will undoubtedly be placed in one of the town's museums.

Washington takes pride in what its local community has done; it has renovated many of the early log cabins and brick buildings, their nearby grounds, and an herb garden.

Gradually, former residents and other tourists from Kentucky, neighboring Ohio, and points beyond have begun to make visits. They take walking tours; they see old water wells, cabins, and especially the Medfort's Fort cabin (1787), built with planks from the flatboat that brought the earliest pioneers down the Ohio River. They attend the five events the town sponsors: Woodcarver's Day, the Chocolate Festival, the Simon Kenton Festival, Civil War Days, and Frontier Christmas. These are put on by the hard work of local folks, who had no Rockefeller sugar daddy as did Williamsburg, Virginia. It was truly a pioneer venture. Even though its history is imperfect, Old Washington stands out as a model of what can be done by others. The past can live on, with all its wrinkles.

Earth-Friendly Technologies at Work

With promotion of Earth-friendly technologies, many changes similar to those seen in Washington will occur. If such technologies are implemented, these are some characteristics of Appalachia in 2025:

> Electric energy bills are dramatically reduced by the popularization of net metering in all states where it is permitted. Subsidies for solar and wind generation have made these sources competitive and even lower priced than nonrenewable fuel sources. Electric demand holds steady due to more widespread wind and solar applications and energy conservation measures, especially efficient lighting.
>
> Food costs have leveled off because of the increased popularity of gardening in the region and an abundance of fresh organic produce now available in stores and the growing numbers of farmers' markets. More homegrown food

means more money stays within the region and is not drained away by grocery store chains and fast food companies. Obesity levels have declined, due in part to emphasis on exercise for youth, availability of better food, improved school menus, and educational efforts by civic and health-care groups.

The forests are being restored through better silviculture practices. Federal laws require licenses and specify conditions for ginseng diggers and other wildcrafters. Reforestation efforts have started to reduce the severity of flooding in the region.

While land is still being developed for commercial purposes, there is a growing awareness that the scenic potential of the land is an asset worth preserving. As a result, tourism has increased. Bed and breakfasts and retreat facilities provide a steady income for numerous property holders in scenic areas. With a decline in surface coal extraction, less land is in need of reclamation.

A decline in the proportion of manufactured and mobile homes increases the region's tax base and allows for more funds for schools and other county needs. Buildings constructed of local materials boost sustainable use of native wood resources while reducing costs.

Transportation, while still a problem in some areas, has improved somewhat. Fewer coal trucks are on the highways than two decades ago, making the roads safer and more easily maintained. Federally subsidized public trains running along established routes are a boon, especially for senior citizens who find it hard to drive. A majority of newer automobiles are fuel-efficient hybrids, solar electric, or fuel cell; hydrogen filling stations are now found in most major towns in the region. Some efforts are being made to develop a regional hiking and bicycling network. All-terrain vehicles are licensed and ridership is limited to those over sixteen years old.

Waste is being better managed through recycling programs in every county and mandatory garbage pickup systems. Salvaging centers for building materials are now in major marketing areas.

Water quality is still a major concern but some improvement is seen. Utilities are owned by local communities. Auxiliary water systems are now in use for watering gardens, supplying drinking water for pets, washing cars, doing laundry, and flushing some toilet systems. A popular farm pond program is in place; some ponds are used for fee fishing and for irrigation and other purposes.

APPALACHIA WITH A NEW VISION

Appalachian Earth-friendly technologies raise issues dealing with broader economic, ecological, social, community, and psychological relationships. One region, even if transformed for efficiency and sustainability, cannot change the entire world—but it could be an important and energizing start. Appalachia enjoyed much self-reliance in earlier days, both before and in the early days of European settlement. It could happen again: this self-reliance in the age of the Internet could prove to be a catalyst for the rest of the world.

Earth-friendly technologies present an amazing array of promises for any community in any bioregion on the planet. If these technologies are adopted, they will be liberating for both the individual user and the community. They need to be affordable, environmentally benign, user-friendly, and community-building by their very nature; they must not tax the resources of people. The users of the technologies are open to the advice of experts about their limits, advantages, and applications but are not beholden to an elite, for appropriate technology involves a dialogue between users and experts. Users of appropriate technology are far removed from the vulnerabilities of complex, highly computerized technologies or those tied to multiple systems of service, maintenance, and commerce.

All people can be appropriate technologists, for these applications are universal in scope, spiritual in depth, and most conducive to cooperation at every level. A friendly appropriate technology is truly a goal worth pursuing, a tool worth using, and a reason for hope for a safe, sustainable, ecological, free, and robust relationship between nature and human culture. And this is as true for Appalachia as it is for every bioregion across this beautiful blue-green planet we call home.

Postscript: Communications

We as a people acknowledge and appreciate the positive benefits of modern technology that has created computers and useful modern communications systems. Progress has always been a goal for our non-native people. But with progress there is greed and competition. If continued, this progress will destroy our Mother Earth and its Creation. Let us not wait any longer. Let us work with common ground and begin to work together to do what can be done to conserve and protect our lives, our children's lives, and all life. So Mother Earth can begin to heal, as we heal, together with one mind, one heart, one Spirit and one Prayer.

William Commanda, keeper of the sacred wampum belts of the Algonquin Nation

Is it right to talk about friendly or appropriate technology trends in the twenty-first century without discussing communication technologies? These technologies include the cell phone, the personal computer, and the Internet, which have dramatically altered the present-day world. More than any others, these three have made a profound impact on the movement toward globalization, for better or worse. These technologies have allowed people in developing countries to bypass the long process of industrialization and creation of an expensive communication infrastructure to become more nearly equal partners in access to general information.

HISTORY OF COMMUNICATION

The history of technology is a history of communication. It is not a history of constant improvement, for progress has depended on many factors, such as the ability of governments to ensure the safe and relatively rapid transfer of information. In Roman times the postal system was highly efficient: good roads and the Pax Romana allowed for the free flow of correspondence. But after the Roman Empire's fall and the coming of the Dark Ages, communications became more difficult and the free flow of information was no longer guaranteed.

Over a longer period of time, other improvements in communication occurred: faster ships, railroads, the telegraph, the telephone, undersea cables, motorized transport, radio, television, satellites, computers, and cell towers.

While older forms of communication could not match the speed of modern electronics, news and knowledge traveled faster than one might expect. For example, I belong to the Society of Jesus; our founder, Saint Ignatius Loyola, impressed upon his company the need to become mobile, to travel to all parts of the world, and to still keep a companionship of quality based on letters, conferences, and the *Relations,* or annual travel and mission reports. In 1712 a Jesuit in Asia wrote in his annual report about Asian ginseng *(Panax ginseng).* Within a mere three years a Jesuit in Canada had read the report and discovered the plant's cousin *Panax quinquefolium,* and the first export of the ginseng root from North America through France to China had taken place.

The recent widespread access to information is a global revolution, allowing a more intense sharing of ideas, not solely the shipping of bulky materials. The challenge now is to guarantee the secure delivery of information.

In World War II and its aftermath, radio brought both news and hope to people living under totalitarian regimes, and even in the age of the Internet, the radio remains an important communications tool. We are avid nocturnal BBC listeners and have come to regard its *Outlook* program, which takes calls from people on every continent, as a glue that will bring our world together. Of course, these callers must have a general proficiency in English, but the program offers a global experience that previous centuries could hardly dream about. Radio can help cement the aspirations of good folks worldwide who strive for health, education, social justice, and tolerance. The transmission devices are sophisticated, but the programs are received through a rather simple, low-cost electronic device and can be heard by millions when not deliberately jammed by tyrants.

Are these modern forms of communication appropriate technology? Certainly the systems required for their functioning are far more sophisticated than an efficient hoe or cooking stove. But producing solar photovoltaic cells is also a sophisticated process, and some other technologies also take extra effort and planning. Appropriateness may be judged by the amount of resources required to sustain these technologies, the number of persons benefiting from them, and the cost and access for each user.

Electronic forms of communication can be efficient in themselves—for example, the cell phone or the Internet can save travel time and expenses. But some people become overly dependent upon television and other electronic media. Many people watch four to six hours of television a day. Television may help standardize language, furnish entertainment, and provide up-to-date news coverage, but often content is limited and viewers are propagandized in quite subtle ways.

Let us pay special attention to two devices: the cell phone and the personal computer (and along with that, Internet service).

CELL PHONES

Cell phones are somewhat affordable, although more expensive and less reliable than land-based phones. Many people benefit from instant communication. The cost of cell phones is low enough that millions of even lower-income people in Africa and China now have these devices; in 2005, 45 percent of cell phone sales were for the Chinese market.

Cell phones have benefits beyond their affordability. Because they are mobile, they can be used to report emergencies outside the home or summon medical or police help, resulting in lives saved. They can allay fears for those wanting to know where a loved one is at a given time; and thanks to GPS (global positioning systems), they can be used to determine the whereabouts of individuals (although privacy problems may soon arise through overuse). They can save travel by making it possible to ask a spouse or friend who is more conveniently located to fetch a product or run an errand. The cell phone becomes a companion to those in unfamiliar or frightening surroundings, because someone else is at least remotely available to help in time of need.

Like all technologies, even appropriate ones, cell phones also have limitations and can be misused. One of the major drawbacks of cell phones is excessive or inappropriate use: people will talk on cell phones everywhere, from the movies to restaurants to public transportation. The degree to which the use of cell phones while driving causes accidents is being debated, and cell phone use while driving is regulated in some places. Can the overuse of cell phones lead to compulsive behavior, the need to be talking with others at all times, and a decrease in and the destruction of silent space—the time for reflection? Why should others intrude on my private space while I'm waiting for or riding in a plane or bus?

Conversations are often interrupted by a ringing cell phone, and an added burden results when cell phone users regard any message as more urgent than the current conversation. In public places the self-importance and loud voice of the person talking on a cell phone becomes a not-so-subtle form of noise pollution. The theory that electromagnetic radiation from a cell phone held to the ear will cause physical harm, even cancer, is still under investigation.

Another drawback to cell phones is that coverage in some areas is spotty. It would be economically prohibitive to put cell phone towers in every cove and valley to ensure coverage in remote areas or in mountainous areas like Appalachia. And economics aside, cell phone towers are ugly. They mar pristine areas and cause scenic pollution. However, this type of pollution is not of the order of the damage caused by the removal of a swath of forest cover to make way for utility lines. Some estimate that the need for large numbers of cell towers may be reduced through future technological improvements. In the meantime, some communities are quite disturbed by the towers, though they prove lucrative to the property holders, who receive substantial fees for allowing the use of their property for tower placement.

PERSONAL COMPUTERS AND THE INTERNET

This book was written using the personal computer for word processing and the Internet for additional information. We would never think of returning to writing a book in longhand or typing it on a typewriter. The use of computers in writing is a great time-saver, reducing composition and editing time and allowing for the storage of information and correction of text. A host of writers have come to depend on personal computers and are intertwined with the Internet. These forms of communication do prove suitable and appropriate, but they are not without some negative aspects.

Computers are widely accessible to those with some degree of training, even in largely nonindustrialized countries. In developing countries, a single computer in a village can be used by a number of individuals with basic education, and the Internet is open to them twenty-four hours a day. Their window to the outside world allows them to bypass dominant overlords who would otherwise dictate what information they would receive. Access to the Internet is changing the political field and helping to end one-party rule and control of resources. The Internet is becoming a place to sound international alarms, solicit needed funding, and bring about the overthrow of oppressive regimes. And it is building a reasonably good track record for just such activities.

A remote Indian villager with access to the Internet today can obtain more information faster than I could have in 1975 when I was living in Washington, D.C., two blocks from the Library of Congress. Admittedly their reference materials are more abbreviated than those available through research in the Library of Congress, but it is still an impressive wealth of information.

Information on the Internet does have some limitations: not all information is reliable, and in many cases material on the Web is limited to relatively recent documents. For instance, the National Technical Information Service (NTIS) lists on the Web only documents from 1990 on.

Computers also have other limitations. The 1996 book *Silicon Snake Oil* discussed a number of problems with excessive computer dependence or misuse. Many of the predictions have proved true, for what is appropriate can also become inappropriate through misuse.

The fragile nature of sophisticated technologies has been mentioned many times. When technologies are interconnected, the overall vulnerability of the entire system is compounded. A national electric grid is complex and vulnerable; a computerized electric grid is doubly vulnerable. Individual homes linked through an electric grid and equipped with net metering procedures are also more vulnerable than a home with an independent solar system, even if the home has its own computer-controlled systems. In an interconnected system, things could go wrong and affect large numbers of linked units and do so quite suddenly.

Hackers or budding terrorists create computer viruses that, through the Internet, can infect millions of computers, spreading havoc. This threat then requires the development of more sophisticated and expensive safeguards, which in turn must be installed, maintained, and updated. Some regard this as making the personal computer far less people-friendly than originally intended. The real possibility of a terrorist attack by a rather sophisticated hacker is always present.

Rumors and lies also spread easily on the Internet. It takes little to blacken a reputation. Once a comment is on the Internet, it can remain as long as there are Internet search engines. The potential damage is immense.

Computer use also raises safety issues. People are spending too much time at a computer screen these days. Some medical experts say that time at the computer should be limited to two hours a day, but that is laughable to millions. A number of physical problems could arise, including neck and back strain, carpal tunnel syndrome, and eye problems. Many of the computer savvy spend hours tapping into a wide variety of entertainment opportunities. While this can be a healthy occupation, parents often complain that computer games take children away from social activity or outdoor sunlight, fresh air, and exercise. Adults can suffer as well in various ways from misuse of computer work or free time.

A growing number of people feel uneasy about e-mail. It is just too popular. We feel we are drowning in e-mail. E-mail does have its advantages: it is less expensive than surface mail; it is fast and allows a fast response. We keep in touch with people with whom we would not otherwise communicate; we turn their electronic messages into hard copy by a simple stroke of the mouse; and we reply in an instant. However, amid all the e-mail benefits several questions are raised. Are important communications sometimes submerged in a sea of spam? Is enough time given to thought before we hit the Send button—especially in emotionally charged situations? Are e-mail messages given the same attention as other mail? Do we really need to receive messages at all hours, even during our rest and vacation time? Doesn't the culture of e-mail take up more time rather than reduce communication time? Does it adversely affect grammar? Why all the chatter?

In spite of these critiques and questions, it must be said that e-mail enhances communication and, if used mindfully, can help facilitate our informed transition to more Earth-friendly technologies.

Computers can encourage a certain intellectual laziness. A friend was tutoring and reprimanded his student for his poor spelling record. The kid replied, "I've got a spell checker for that." Plagiarism is a growing problem, as it is easy to take someone else's work. Another friend who taught basic English said it required being a detective looking for key words or phrases as signals of theft.

Computer models change fairly rapidly as technology advances—a form of planned obsolescence. As some older computer equipment goes out of fashion, the retention of computerized information becomes more and more difficult. The permanence of these older records is sometime tenuous and worries storage specialists. Records are no longer written on parchment and may easily disappear.

The computers themselves, however, don't disappear. They are hard to dispose of. Efforts have been made to reduce toxic materials used in the manufacture of computer equipment, but little effort has gone into providing for disposal of older computers. Often the outer cases of computers could be reused and some monitors recycled, but there still is a growing solid waste problem.

Resources

CHAPTER 1 SOLAR PHOTOVOLTAICS

Appalachia—Science in the Public Interest (ASPI). *Rural Solar Photovoltaic Use.* Technical Paper 14. Mount Vernon, Ky.: ASPI Publications, 1992.

Berman, Daniel M., and John T. O'Connor. Foreword by Ralph Nader. *Who Owns the Sun? People, Politics, and the Struggle for a Solar Economy.* White River Junction, Vt.: Chelsea Green, 1996.

Bills, Joshua. *ASPI Solar House.* Technical Paper 40. Mount Vernon, Ky.: ASPI Publications, 1997.

———. *Solar Photovoltaic Advances: Home-Grown Electricity and Electric Utility Grid.* Technical Paper 49. Mount Vernon, Ky.: ASPI Publications, 1999.

Davidson, Joel, and Fran Orner. *New Solar Electric Home: The Photovoltaics How-To Handbook.* 3rd ed. Christchurch, New Zealand: Stonefield Publishing, 2005.

Endecon Engineering with Regional Economic Research. *A Guide to Photovoltaic (PV) System Design and Installation.* Publication No. 500-01-020. Sacramento: California Energy Commission, 2001.

Home Power Magazine, PO Box 520, Ashland, Ore. 97520; http://www.homepower.com.

Komp, Richard J. *Practical Photovoltaics: Electricity from Solar Cells.* 3rd ed. Ann Arbor, Mich.: Aatec Publications, 2001.

Schaeffer, John, ed. *Real Goods Solar Living Sourcebook: The Complete Guide to Renewable Energy Technologies and Sustainable Living.* 12th ed. Gabriola Island, B.C.: New Society Publishers, 2005.

Solar Energy International. *Photovoltaics: A Design and Installation Manual.* Gabriola Island, B.C.: New Society Publishers, 2004.

Strong, Steven J., with William G. Scheller. *The Solar Electric House: Energy for the Environmentally-Responsive, Energy-Independent Home.*

Still River, Mass.: Sustainability Press, 1994. Distributed by Chelsea Green.

Yago, Jeffrey R. *Achieving Energy Independence—One Step at a Time.* Gum Spring, Va.: Dunimis Technology, 1999.

CHAPTER 2 **MICROHYDROPOWER**

Alward, Ron, Sherry Eisenbart, and John Wolkman. *Micro-HydroPower: Reviewing an Old Concept.* Butte, Mont.: National Center for Appropriate Technology (NCAT), 1979. (National Center for Appropriate Technology, PO Box 3838, Butte, Mont. 59702; 406-494-4572.)

Brown, Norman, ed. *Renewable Energy Resources and Rural Applications in the Developing World.* Boulder, Colo.: Westview Press, in association with the American Association for the Advancement of Science, 1978.

Chiras, Dan. *The Homeowner's Guide to Renewable Energy: Achieving Energy Independence through Solar, Wind, Biomass and Hydropower.* Gabriola Island, B.C.: New Society Publishers, 2006.

Curtis, Dan, et al. *Going with the Flow: Small-Scale Water Power Made Simple.* Machynlleth, Powys, UK: Centre for Alternative Technology, 1999.

Davis, Scott. *Microhydro: Clean Power from Water.* Gabriola Island, B.C.: New Society Publishers, 2003.

EG&G Idaho. *Microhydropower Handbook.* 2 vols. U.S. Department of Energy, 1983. National Technical Information Service (NTIS) DE83-006697 and DE83-006698.

Hamm, Hans W. *Low-Cost Development of Small Water Power Sites.* Arlington, Va.: Volunteers in Technical Assistance (VITA), 1971 and 1975.

Klunne, Wim Jonker. http://www.microhydropower.net. [A Web portal.]

Leckie, J., et al. *More Other Homes and Garbage: Designs for Self-Sufficient Living.* San Francisco: Sierra Club Books, 1981.

McGuigan, Dermot. *Harnessing Water Power for Home Energy.* Charlotte, Vt.: Garden Way Publishing, 1978.

Village Earth: The Consortium for Sustainable Village-Based Development, PO Box 797, Fort Collins, Colo. 80522; 970-491-5754; http://www.villageearth.org.

Warren, John, and Paul Gallimore. *Guide to Development of Small Hydroelectric and Microhydroelectric Projects in North Carolina.* Research Triangle Park, N.C.: North Carolina Alternative Energy Corporation, 1983.

CHAPTER 3 **WIND POWER**

Ewing, Rex A. *Power with Nature: Solar and Wind Energy Demystified.* Masonville, Colo.: Pixyjack Press, 2003.

Gipe, Paul. *Wind Energy Basics: A Guide to Small and Micro Wind Systems.* White River Junction, Vt.: Chelsea Green, 1999.

———. *Wind Power: Renewable Energy for Home, Farm, and Business.* White River Junction, Vt.: Chelsea Green, 2004.

Kemp, William H. *The Renewable Energy Handbook: A Guide to Rural Energy Independence, Off-Grid and Sustainable Living.* Gabriola Island, B.C.: New Society Publishers, 2005.

Piggott, Hugh. *It's a Breeze: A Guide to Choosing Windpower.* New Futures. Machynlleth, Powys, UK: Centre for Alternative Technology, 1995.

———. *Windpower Workshop.* Gabriola Island, B.C.: New Society Publishers, 2001.

CHAPTER 4 **WOOD HEATING**

Aprovecho Research Center, 80574 Hazelton Road, Cottage Grove, Ore. 97424; 541-942-8198; http://www.aprovecho.net.

ASPI. *Contraflow Masonry Heaters/Ovens.* Technical Paper 12. Mount Vernon, Ky.: ASPI Publications, 1991.

Barden, Albert, and Heikki Hyytiainen. *Finnish Fireplaces: Heart of the Home.* Norridgewock, Maine: Maine Wood Heat, 1984.

The Chimney Sweep. *Woodstoves: Will They Be Outlawed? An Overview of EPA Woodstove Emissions Regulations.* Bellingham, Wash.: Chimney Sweep, 1996–2005. Available online at http://www.chimneysweeponline.com/hoepareg.htm.

Energy Efficiency and Renewable Energy Clearinghouse, PO Box 3048, Merrifield, Va. 22116; 800-363-3732; http://www.greenbiz.com.

Gulland, John. "The Art of the Wood Cookstove: Warm Up Your Home, Hearth, and Heart." *Mother Earth News,* December/January 2005.

Hayden, A. C. S. "Fireplaces: Studies in Contrasts." *Home Energy,* n.d. Available online at http://hearth.com/what/more/skip.html.

Lyle, David. *The Book of Masonry Stoves.* White River Junction, Vt.: Chelsea Green, 1996.

Pahl, Greg. *Natural Home Heating: The Complete Guide to Renewable Energy Options.* White River Junction, Vt.: Chelsea Green, 2003.

Ritchie, Ralph W. *All That's Practical about Wood: Stoves, as a Fuel, Heating—Including Wood Pellet Stoves.* 2nd ed. Springfield, Ore.: Ritchie Unlimited Publications, 1998.

Schimmoeller, Mark. *"Elbow Torch" Efficient Wood Cooker.* Technical Paper 39. Mount Vernon, Ky.: ASPI Publications, 1996.

Thomas, Dirk. *The Woodburner's Companion: Practical Ways of Heating with Wood.* Chambersburg, Pa.: Alan C. Hood, 2000.

U.S. Environmental Protection Agency, Consumer Product Safety Commission, and the American Lung Association. *What You Should Know about Combustion Appliances and Indoor Air Pollution.* Available online at http://www.epa.gov/iaq/pubs/combust.html.

Wik, Ole. *Wood Stoves: How to Make and Use Them.* Portland, Ore.: Alaska Northwest Books, 1977.

CHAPTER 5 **SOLAR HEATING APPLICATIONS**

ASPI. *Breadbox Solar Water Heaters.* Technical Paper 9. Mount Vernon, Ky.: ASPI Publications, 1991.

———. *The Lilongwe Solar Box Cooker.* Technical Paper 67. Mount Vernon, Ky.: ASPI Publications, 2003.

———. *Quilted Insulated Shades.* Technical Paper 21. Mount Vernon, Ky.: ASPI Publications, 1993.

———. *Solar Box Cooker.* Technical Paper 1. Mount Vernon, Ky.: ASPI Publications, 1990.

Lane, Thomas. *Solar Hot Water Systems—1977 to Today: Lessons Learned.* Gainesville, Fla.: Energy Conservation Services of North Florida, (http://www.ecs.solar.com), 2003.

Langdon, William K. *Movable Insulation.* Emmaus, Pa.: Rodale Press, 1980.

National Center for Appropriate Technology. *Breadbox Solar Water Systems.* Butte, Mont.: NCAT, 1981.

Radabaugh, Joseph. *Heaven's Flame: A Guide to Solar Cookers.* Ashland, Ore.: Home Power, 1998.

Rocky Mountain Institute. *Cleaning Appliances,* Home Energy Brief 6. Snowmount, CO: Rocky Mountain Institute, 1995. Available at http://www.rmi.org/images/other/Energy/E04-16_HEB6CleaningApps.pdf.

Shurcliff, William A. *Thermal Shutters and Shades.* Andover, Mass.: Brick House Publishing, 1980.

Stanley, Tomm. *Going Solar: Understanding and Using the Warmth in Sunlight.* Christchurch, New Zealand: Stonefield Publishing, 2005. Distributed by Chelsea Green.

Volunteers in Technical Assistance. *Solar Hot Water Heater.* Mount Rainier, Md.: VITA, n.d.

———. *Understanding Solar Cookers and Ovens.* Arlington, Va.: VITA, n.d.

CHAPTER 6 **SHADE TREES AND WINDBREAKS**

Foster, Ruth S. *Landscaping That Saves Energy and Dollars.* Guilford, Conn.: Globe Pequot, 1994.

Girgis, Magdy. *Landscaping for Energy Conservation.* Energy Note / Florida Solar Energy Center. Cocoa, Fla.: Florida Solar Energy Center, 1985.

Glickman, Marshall. "Saving Energy and Money with Landscaping." *Consumers' Research Magazine,* September 1994.

Hill, Lewis. *Fruits and Berries for the Home Garden.* North Adams, Mass.: Storey, 1992.

Kourik, Robert, and Jeff Jackson. "Made in the Shade." *Garbage,* June-July 1993, pp. 44 ff.

Moffat, Anne Simon, and Mark Schiler. *Energy-Efficient and Environmental Landscaping: Cut Your Utility Bills by Up to 30 Percent and Create a Natural Healthy Yard.* South Newfane, Vt.: Appropriate Solutions Press, 1994.

Walker, Lloyd R. *Landscaping for Energy Conservation.* Service in Action. Boulder: Colorado State University Extension Service, 1992. Available online at http://www.ext.colostate.edu/pubs/garden/07225.html.

Welch, William C. *Landscaping for Energy Conservation.* Texas A & M University System, Extension Horticulture Information Resource, n.d. Available at http://aggie-horticulture.tamu.edu/extension/homelandscape/energy/energy.html.

CHAPTER 7 FOOD PRESERVATION

ASPI. *Root Cellars.* Technical Paper 24. Mount Vernon, Ky.: ASPI Publications, 1993.

———. *Solar Food Dryer* [construction instructions]. Technical Paper 6. Mount Vernon, Ky.: ASPI Publications, 1991.

Bailey, Roberta R. "Learn to Can for Homegrown Flavor." *Mother Earth News,* August/September 2005, 114–20.

Bubel, Mike, and Nancy Bubel. *Root Cellaring: The Simple No-Processing Way to Store Fruits and Vegetables.* Pownal, Vt.: Storey, 1991.

Cavagnaro, David. "How to Store Fresh Vegetables." *Mother Earth News,* December 2004 / January 2005, 46–54.

Clark, Wayne A. *Plans for an Appalachian Solar Food Dryer.* Mount Vernon, Ky.: ASPI Publications, 1980.

Fodor, Eben. *The Solar Food Dryer: How to Make Your Own Low-Cost, High-Performance, Sun-Powered Food Dehydrator.* Gabriola Island, B.C.: New Society Publishers, 2006.

Hupping, Carol, and the staff of the Rodale Food Center. *Stocking Up.* 3rd ed. New York: Simon & Schuster, 1990.

Maxwell, Steve. "Build a Basement Root Cellar." *Mother Earth News,* December 2004 / January 2005, 55–56.

Rhodes, Phillip, and Brian Leatart. "Better than Fresh." *Natural Health,* April 2005, 62–107.

Warren, Piers. *How to Store Your Garden Produce.* Totnes Devon, UK: Green Books, 2003. Distributed by Chelsea Green.

CHAPTER 8 EDIBLE LANDSCAPING

Creasy, Rosalind. *The Complete Book of Edible Landscaping: Home Landscaping with Food-Bearing Plants and Resource-Saving Techniques.* San Francisco: Sierra Club Books, 1982.

———. *The Edible Flower Garden.* Edible Garden Series. London: Periplus Editions, 1999.

Frisch, Karl von. *The Dance Language and Orientation of Bees.* Trans. Leigh E. Chadwick. Cambridge, Mass.: Belknap Press / Harvard University Press, 1967.

Gardner, Jigs, and Jo Ann Gardner. *Gardens of Use and Delight: Uniting the Practical and Beautiful in an Integrated Landscape.* Golden, Colo.: Fulcrum Publishing, 2002.

Gardner, Jo Ann. *Living with Herbs: A Treasury of Useful Plants for the Home and Garden.* Woodstock, Vt.: Countryman Press, 1997.

Gessert, Kate Rogers. *The Beautiful Food Garden: Creative Landscaping with Vegetables, Herbs, Fruits, and Flowers.* Charlotte, Vt.: Garden Way Publishing, 1987.

Hagy, Fred. *Landscaping with Fruit and Vegetables.* Woodstock, N.Y.: Overlook Hardcover, 2001.

Kourik, Robert. *Designing and Maintaining Your Edible Landscape Naturally.* East Meon, Hampshire, UK: Permanent Publications, 2005. Distributed by Chelsea Green.

Muhly, Ernie. *Pollination.* Technical Paper 31. Mount Vernon, Ky.: ASPI Publications, 1996.

CHAPTER 9 **INTENSIVE AND ORGANIC GARDENING AND ORCHARDING**

ASPI. *Establishing a Floral-Vegetable Garden.* Technical Paper 53. Mount Vernon, Ky.: ASPI Publications, 1999.

———. *How Much Do You Mulch?* Technical Paper 22. Mount Vernon, Ky.: ASPI Publications, 1993.

———. *Organic and Intensive Gardening.* Technical Paper 13. Mount Vernon, Ky.: ASPI Publications, 1992.

Bradley, Fern Marshall, and Barbara W. Ellis, eds. *Rodale's All-New Encyclopedia of Organic Gardening.* Emmaus, Pa.: Rodale Press, 1993.

Coleman, Eliot. *The New Organic Grower: A Master's Manual of Tools and Techniques for the Home and Market Gardener.* 2nd ed. White River Junction, Vt.: Chelsea Green, 1995.

Ellis, Barbara W., and Fern Marshall Bradley. *The Organic Gardener's Handbook of Natural Insect and Disease Control: A Complete Problem-Solving Guide to Keeping Your Garden and Yard Healthy without Chemicals.* Rev. ed. Emmaus, Pa.: Rodale Press, 1996.

Fritsch, Albert J. *Backyard Gardens: The Last Frontier.* Technical Paper 46. Mount Vernon, Ky.: ASPI Publications, 1998.

HortIdeas, 750 Black Lick Road, Gravel Switch, Ky. 40328.

Jeavons, John. *How to Grow More Vegetables: And Fruits, Nuts, Berries, Grains, and Other Crops.* 6th ed. Berkeley, Calif.: Ten Speed Press, 2002.

Logsdon, Gen. *Organic Orcharding.* Emmaus, Pa.: Rodale Press, 1981.

Organic Gardening. Rodale Press, Emmaus, Pa.

Pest Management Supply; 311 River Drive, Hadley, Mass. 01035; 800-272-7672.

Phillips, Michael. *The Apple Grower: A Guide for the Organic Orchardist.* White River Junction, Vt.: Chelsea Green, 2005.

Rincon-Vitova Insectaries, 108 Orchard Drive, Ventura, Calif. 93001; 800-248-2847; http://www.rinconvitova.com.

CHAPTER 10 **REGIONAL HERITAGE PLANTS**

Adams, Denise Wiles. *Restoring American Gardens: An Encyclopedia of Heirloom Ornamental Plants, 1640–1940.* Portland, Ore.: Timber Press, 2004.

The American Chestnut Cooperators' Foundation, 2667 Forest Service Road 708, Newport, Va. 24128; http://ipm.ppws.vt.edu/griffin/accf.html.

The American Chestnut Foundation, 469 Main Street, Suite 1, PO Box 4044, Bennington, Vt. 05201; 802-447-0010; http://www.acf.org.

Anagnostakis, Sandra L. Connecticut Agricultural Experiment Station, Box 1106, New Haven, Conn. 06504-1106.

Appalachian Heirloom Seed Conservancy, PO Box 519, Richmond, Ky. 40476; kentuckyseeds@hotmail.com.

ASPI. *Herb Growing in Appalachia.* Technical Paper 62. Mount Vernon, Ky.: ASPI Publications, 2002.

———. *Seed Saving.* Technical Paper 29. Mount Vernon, Ky.: ASPI Publications, 1994.

Browning, Frank. *Apples.* New York: North Point Press, 1998.

Bryant, Geoff. *Propagation Handbook: Basic Techniques for Gardeners.* Mechanicsburg, Pa.: Stackpole Books, 1995.

Calhoun, Creighton Lee, Jr. *Old Southern Apples.* Blacksburg, Va.: McDonald & Woodward, 1995.

Creasy, Rosalind. *The Edible Heirloom Garden.* Boston: Periplus Editions, 1999.

Gallimore, Paul. *American Chestnut Restoration.* Technical Paper 44. Mount Vernon, Ky.: ASPI Publications, 1998.

———. *Heirloom Apples in Central Appalachia.* Technical Paper 50. Mount Vernon, Ky.: ASPI Publications, 1999.

Seed Savers Exchange; Flower and Herb Exchange, 3094 North Winn Road, Decorah, Iowa 52101; 563-382-5990; http://seedsavers.org/Home.asp.

Seeds of Change, PO Box 15700, Santa Fe, N.M. 87506-5700; 888-762-7333; http://www.seedsofchange.com.

Southern Exposure Seed Exchange, PO Box 460, Mineral, Va. 23117; 540-894-9480; http://www.southernexposure.com.

Watson, Benjamin J. *Taylor's Guide to Heirloom Vegetables: A Complete Guide to the Best Historic and Ethnic Varieties.* Boston: Houghton Mifflin, 1996.

CHAPTER 11 SOLAR GREENHOUSES AND SEASON EXTENDERS

Alward, Ron, and Andy Shapiro. *Low-Cost Passive Solar Greenhouses: A Design and Construction Guide.* 2nd ed. Butte, Mont.: National Center for Appropriate Technology, 1980.

ASPI. *Solar Greenhouses.* Technical Paper 4. Mount Vernon, Ky.: ASPI Publications, 1990.

———. *Year-Round Gardening.* Technical Paper 7. Mount Vernon, Ky.: ASPI Publications, 1991.

Bellows, Barbara. *Solar Greenhouses: Horticulture Resource List.* Fayetteville, Ark.: Appropriate Technology Transfer for Rural Areas–National Sustainable Agriculture Information Service, 2003. ATTRA–National Sustainable Agriculture Information Service, PO Box 3657, Fayetteville, Ark. 72702; 800-346-9140.

Claude Terry & Associates. *Sunspaces for the Southeastern United States: Principles and Construction of a Passive Solar Greenhouse.* Atlanta.: U.S. Department of Energy, Region 4, Office of Appropriate Technology, n.d.

Coleman, Eliot. *Four-Season Harvest: Organic Vegetables from Your Garden All Year Long.* White River Junction, Vt.: Chelsea Green, 1992.

Elliott, Brook. "Using Wire Mesh in the Garden." *Mother Earth News,* June/July 2002, 96 ff.

Lewis, Hill. *Cold-Climate Gardening: How to Extend Your Growing Season by at Least 30 Days.* North Adams, Mass.: Storey, 1987.

Poisson, Leandre. *Solar Gardening: Growing Vegetables Year-Round the American Intensive Way.* White River Junction, Vt.: Chelsea Green, 1994.

Relf, Diane. *Season Extenders.* Rev. ed. Blacksburg: Virginia Cooperative Extension, 2000.

Stetson, Fred. *Extend Your Garden Season: Row Covers and Mulches.* Storey Country Wisdom Bulletin A-148. North Adams, Mass.: Storey, 1996.

CHAPTER 12 **WILDLIFE HABITAT RESTORATION**

ASPI. *The ASPI Nature Center Educational Program.* Technical Paper 60. Mount Vernon, Ky.: ASPI Publications, 2000.

———. *Coexisting with Wildlife.* Technical Paper 17. Mount Vernon, Ky.: ASPI Publications, 1992.

———. *Domestic Wildscape.* Technical Paper 48. Mount Vernon, Ky.: ASPI Publications, 1998.

———. *Nature Trails.* Technical Paper 10. Mount Vernon, Ky.: ASPI Publications, 1991

Evans, Brent, and Carolyn Chipman-Evans. *How to Create and Nurture a Nature Center in Your Community.* Austin: University of Texas Press, 1998.

Evans, Julian. *A Wood of Our Own.* East Meon, Hampshire, UK: Permanent Publications, 2005. Distributed by Chelsea Green.

Federal Interagency Stream Restoration Working Group. *Stream Corridor Restoration: Principles, Processes and Practices.* Washington, D.C.: Natural Resources Conservation Service, 1998. Available online at http://purl.access.gpo.gov/GPO/LPS63051.

Imhoff, Dan. *Farming with the Wild: Enhancing Biodiversity on Farms and Ranches.* San Francisco: Sierra Club Books, 2003.

Interagency Workgroup on Wetland Restoration. *An Introduction and User's Guide to Wetland Restoration, Creation, and Enhancement.* Available online at http://www.epa.gov/owow/wetlands/restore/finalinfo.html.

Jackson, Dana L., and Laura L. Jackson, eds. *Farm as Natural Habitat: Reconnecting Food Systems with Ecosystems.* Washington, D.C.: Island Press, 2002.

Loewer, Peter. *The Wild Gardener.* Mechanicsburg, Pa.: Stackpole Books, 1991.

Maehr, David S., Reed F. Noss, and Jeffery L. Larkin, eds. *Large Mammal Restoration: Ecological and Sociological Challenges in the 21st Century.* Washington, D.C.: Island Press, 2001.

Merilees, Bill, and William J. Merilees. *Attracting Backyard Wildlife: A Guide for Nature Lovers.* Stillwater, Minn.: Voyageur Press, 2002.

Mizejewski, David. *NWF's Attracting Birds, Butterflies, and Other Backyard Wildlife.* Upper Saddle River, N.J.: Creative Homeowner Press, 2003.

Sauer, Leslie Jones. *The Once and Future Forest: A Guide to Forest Restoration Strategies.* Washington, D.C.: Island Press, 1998.

Wilhelm, Gene. *Appalachian Highlands: A Field Guide to Ecology.* Mount Vernon, Ky.: ASPI Publications, 1996.

———. *Through the Seasons: A Celebration of Creation.* Oikos Publications, 1990. Distributed by ASPI Publications.

CHAPTER 13 **NONTIMBER FOREST PRODUCTS**

ASPI. *Alternative Forest Products.* Technical Paper 28. Mount Vernon, Ky.: ASPI Publications, 1994.

———. *Guide to Use of Non-Timber Products.* Technical Paper 42. Mount Vernon, Ky.: ASPI Publications, 1998.

Cole, David. *Ginseng in Appalachia.* Technical Paper 38. Mount Vernon, Ky.: ASPI Publications, 1996.

Hammett, A. L. "Greenery: An Opportunity for Forest Landowners." *Forest Landowner,* March/April 2002, 44–46.

Institute for Culture and Ecology, http://www.ifcae.org/ntfp/ with bibliographic and product databases for nontimber forest products.

Jones, Eric T., et al. *Nontimber Forest Products in the United States.* Lawrence: University Press of Kansas, 2002.

Muhly, Ernie. *Mushroom Culturing in Appalachia.* Technical Paper 36. Mount Vernon, Ky.: ASPI Publications, 1996.

National Network of Forest Practitioners, 305 South Main Street, Providence, R.I. 02903; 401-273-6507; http://www.nnfp.org.

Persons, W. Scott, and Jeanine M. Davis. *Growing and Marketing Ginseng, Goldenseal, and Other Woodland Medicinals.* Fairview, N.C.: Bright Mountain Books, 2005.

Pritts, Kim. *Ginseng: How to Find, Grow, and Use America's Forest Gold.* Mechanicsburg, Pa.: Stackpole Books, 1995.

Williams, Greg. *Moratorium on Wild American Ginseng Exports.* Technical Paper 57. Mount Vernon, Ky.: ASPI Publications, 2000.

CHAPTER 14 **SILVICULTURAL PRACTICES**

Anderson, Gary. *Sustainable Logging and Lumber Production.* Technical Paper 35. Mount Vernon, Ky.: ASPI Publications, 1996.

ASPI. *Forest Preservation Techniques.* Technical Paper 16. Mount Vernon, Ky.: ASPI Publications, 1991.

Drengson, Alan, and Duncan Taylor, eds. *Ecoforestry: The Art and Science of Sustainable Forest Use.* Gabriola Island, B.C.: New Society Publishers, 1997.

Foster, Bryan C. *Wild Logging: A Guide to Environmentally and Economically Sustainable Forestry.* Signal Mountain, Tenn.: Mountain Press Publishing, 2002.

Hammond, Herb. *Seeing the Forest among the Trees: The Case for Wholistic Forest Use.* Vancouver, B.C.: Polestar Calendars, 1992.

Hilts, Stewart, and Peter Mitchell. *The Woodlot Management Handbook: Making the Most of Your Wooded Property for Conservation, Income, or Both.* Richmond Hill, Ont.: Firefly Books, 1999.

Israel, David L. "Small Woodlot Management." *Mother Earth News,* February/March 1995, pp. 68 ff.

Kalisz, Paul. *The Practice of Ecoforestry.* Technical Papers 32, 33, and 34. Mount Vernon, Ky.: ASPI Publications, 1996.

Lansky, Mitch, ed. *Forestry as If the Future Mattered.* Hallowell, Maine: Maine Environmental Policy Institute, 2002.

McEvoy, Thom J. *A Sustainable Approach to Managing Woodlands.* Washington, D.C.: Island Press, 2004.

Morsbach, Hans. *Common Sense Forestry.* White River Junction, Vt.: Chelsea Green, 2002.

Southern Appalachian Man and the Biosphere. *Southern Appalachian Assessment Terrestrial Technical Report.* Report 5 of 5. Atlanta: U.S. Department of Agriculture, Forest Service, Southern Region, 1996. Available at http://samab.org/saa/reports/terrestrial/terrestrial.html.

CHAPTER 15 **WILDCRAFTING**

ASPI. *Appalachian Edible Wild Plants.* Technical Paper 23. Mount Vernon, Ky.: ASPI Publications, 1993.

———. *Mushroom Gathering and Culturing in Appalachia.* Technical Paper 36. Mount Vernon, Ky.: ASPI Publications, 1996.

Cech, Richo. *Growing At-Risk Medicinal Herbs: Cultivation, Conservation, and Ecology.* Williams, Ore.: Horizon Herbs, 2002.

Couplan, François. *The Encyclopedia of Edible Plants of North America.* New Canaan, Conn.: Keats Publishing, 1998.

Duke, James A., and Steven Foster. *A Field Guide to Medicinal Plants and Herbs of Eastern and Central North America.* Peterson Field Guides. Boston: Houghton Mifflin, 1999.

Dworkin, Norine. "Where Are All the Flowers Gone?" *Vegetarian Times,* September 1999, 84ff.

Lady Bird Johnson Wildflower Center, 4601 La Crosse Avenue, Austin, Texas 78739; 512-292-4100; http://www.wildflower.org.

North American Mycological Association, 6615 Tudor Court, Gladstone, Ore. 97027-1032; 503-657-7358; http://www.namyco.org.

Peterson, Lee Allen. *A Field Guide to Edible Wild Plants: Eastern / Central North America.* Peterson Field Guide Series. Boston: Houghton Mifflin, 1977.

Phillips, Rodger. *Mushrooms and Other Fungi of North America.* Rev. ed. Richmond Hill, Ont.: Firefly Books, 2005.

Runyon, Linda. *From Crabgrass Muffins to Pine Needle Tea: A National Wild Field Guide.* Shiloh, N.J.: Wild Food, 2002.

Sturdivant, Lee, and Tim Blakeley. *Medicinal Herbs in the Garden, Field, and Marketplace.* Friday Harbor, Wash.: San Juan Naturals, 2003.

United Plant Savers, PO Box 400, East Barre, Vt. 05649, 802-479-9825; http://unitedplantsavers.org. Dedicated to preserving native medicinal plants.

CHAPTER 16 **CONSTRUCTED OR ARTIFICIAL WETLANDS**

Bedwell, Emily. *Compost Toilets and Constructed Wetlands as a Solution for Appalachian Water Problems.* Technical Paper 58. Mount Vernon, Ky.: ASPI Publications, 2000.

Campbell, Craig S., and Michael Ogden. *Constructed Wetlands in the Sustainable Landscape.* Hoboken, N.J.: John Wiley & Sons, 1999.

Kieffer, Jack. *Artificial or Constructed Wetlands.* Technical Paper 30. Mount Vernon, Ky.: ASPI Publications, 1994.

Kusler, Jon A., and Mary E. Kentula. *Wetland Creation and Restoration: The Status of Science.* Washington, D.C.: Island Press, 1990.

Moshiri, Gerald A. *Constructed Wetlands for Water Quality Improvement.* Washington, D.C.: Island Press, 2000.

NutriCycle Systems Graywater Flower Bed. For information, see http://www.nutricyclesystems.com/moreinfo.htm.

U.S. Environmental Protection Agency. *Guiding Principles for Constructed Treatment Wetlands: Providing for Water Quality and Wildlife Habitat.* EPA 843-B-00-003. Available online at http://www.epa.gov/owow/wetlands/watersheds/cwetlands.html.

CHAPTER 17 **LAND RECLAMATION WITH NATIVE SPECIES**

Allen, Amanda. *Exotic Plant Species in Central Appalachia.* Technical Paper 56. Mount Vernon, Ky.: ASPI Publications, 2000.

———. *Kudzu in Appalachia.* Technical Paper 55. Mount Vernon, Ky.: ASPI Publications, 2000.

Blumenstyk, Goldie. "A Scholar Devoted to Kudzu." *Chronicle of Higher Education,* May 16, 1997.

Cullina, William. *Guide to Growing and Propagating Wildflowers of the United States and Canada.* Boston: Houghton Mifflin, 2000.

———. *Native Trees, Shrubs, and Vines: A Guide to Using, Growing, and Propagating North American Woody Plants.* Framingham, Mass.: New England Wildflower Society, 2002.

Kilgo, John C., and John I. Blake. *Ecology and Management of a Forested Landscape: Fifty Years on the Savannah River Site.* Washington, D.C.: Island Press, 2005.

Miller, James H. *Nonnative Invasive Plants of Southern Forests: A Field Guide for Identification and Control.* Rev. ed. General Technical Report SRS-62. Asheville, N.C.: U.S. Department of Agriculture, Forest Service, Southern Research Station, 2003.

Morrison, Michael L. *Wildlife Restoration Techniques for Habitat Analysis and Animal Monitoring: The Science and Practice of Ecological Restoration.* Washington, D.C.: Island Press, 2002.

Sauer, Leslie Jones. *The Once and Future Forest: A Guide to Forest Restoration Strategies.* Washington, D.C.: Island Press, 1998.

Shurtleff, William, and Akiki Aoyagi. *The Book of Kudzu: A Culinary and Healing Guide.* New York: Avery Publishing, 1985.

CHAPTER 18 **RETREAT CABIN SITES**

ASPI. *Hermitages in Appalachia.* Technical Paper 68. Mount Vernon, Ky.: ASPI Publications, 2002.

Fritsch, Albert, and Kristin Johannsen. *Ecotourism in Appalachia: Marketing the Mountains.* Lexington: University Press of Kentucky, 2004.

Mulfinger, David, and Susan E. Davis. *The Cabin: Inspiration for the Classic American Getaway.* Newtown, Conn.: Taunton Press, 2003.

Mulligan, Michael. *Building a Log Cabin Retreat: A Do-It-Yourself Guide.* Boulder, Colo.: Paladin Press, 2002.

Stiles, David. *Rustic Retreats: A Build-It-Yourself Guide.* North Adams, Mass.: Storey, 1998.

U.S. Green Building Council, http://www.usgbc.org, and, for its LEED (Leadership in Energy and Environmental Design) program, http://www.usgbc.org/DisplayPage.aspx?CategoryID=19.

CHAPTER 19 **ENERGY-EFFICIENT PASSIVE SOLAR DESIGN**

Anderson, Bruce, ed. *The Fuel Savers: A Kit of Solar Ideas for Your Home.* Lafayette, Calif.: Morning Sun Press, 1991.

ASPI. *The Solar Demonstration House.* Technical Paper 40. Mount Vernon, Ky.: ASPI Publications, 1997.

Chiras, Daniel D. *The Solar House: Passive Heating and Cooling.* White River Junction: Vt.: Chelsea Green, 2002.

Freeman, Mark. *The Solar Home: How to Design and Build a House You Heat with the Sun.* Mechanicsburg, Pa.: Stackpole Books, 1994.

Hibschman, Dan. *Your Affordable Solar Home.* San Francisco: Sierra Club Books, 1983.

Little, Amanda Griscom. "Super Solar Homes Everyone Can Afford." *Mother Earth News,* December 2004 / January 2005, 34ff.

Mazria, Ed. *The Passive Solar Energy Book: A Complete Guide to Passive Solar Home, Greenhouse, and Building Design.* Emmaus, Pa.: Rodale Press, 1979.

CHAPTER 20 NATURAL COOLING

Arizona Solar Center. "Natural Cooling." In the *Passive Solar Heating and Cooling Design Manual.* Available online at http://www.azsolarcenter.com/design/passive-3.html.

Chiras, Daniel D. *The New Ecological Home: A Complete Guide to Green Building Options.* White River Junction, Vt.: Chelsea Green, 2004.

Eartheasy. "Natural Cooling." Available online at http://eartheasy.com/live_naturalcooling.htm#1a.

Kachadorian, James. *The Passive Solar House: Using Solar Design to Heat and Cool Your Home.* White River Junction, Vt.: Chelsea Green, 1997.

Wilson, Alex, Jennifer Thorne, and John Morrill. *Consumer Guide to Home Energy Savings.* 8th ed. Washington, D.C.: American Council for an Energy-Efficient Economy, 2003.

CHAPTER 21 NATIVE BUILDING MATERIALS

ASPI. *Masonry Rockwork in Appalachia.* Technical Paper 69. Mount Vernon, Ky.: ASPI Publications, 2002.

Chambers, Robert W. *Log Construction Manual: The Ultimate Guide to Building Handcrafted Log Homes.* River Falls, Wis.: Deep Stream Press, 2002.

Earthship Biotecture, PO Box 1041, Taos, N.M. 87571; 505-751-0462; http://www.earthship.org.

Evans, Tanto, and Linda Smiley. *The Hand-Sculpted House: A Practical and Philosophical Guide to Building a Cob Cottage.* White River Junction, Vt.: Chelsea Green, 2002.

Folsom, C. D. "You Can Build with Soil Cement Blocks." *Mother Earth News,* June 1978, 28–31.

Fordice, John. Cob Code Project. Berkeley, Calif. http://www.deatech.com/natural/cobinfo/cobcode.html.

Forestry Suppliers, Inc., 205 West Rankin Street, Box 8397, Jackson, Miss. 39204. Seller of tools.

Hunter, Kaki, and Donald Kiffmeyer. *Earthbag Building: The Tools, Tricks, and Techniques.* Gabriola Island, B.C.: New Society Publishers, 2006.

Kern, Ken. *Owner Builders Guide to Stone Masonry.* New York: Scribner, 1981.

———. *The Owner-Built Home.* New York: Scribner, 1975.

Mackie, B. Allan. *The Owner-Built Log House: Living in Harmony with Your Environment.* Richmond Hill, Ont.: Firefly Books, 2001.

McRaven, Charles. *Building with Stone.* Pownal, Vt.: Storey, 1989.

Price, Thom. *Pressed Earth Blocks: An Inexpensive Building Material.* Technical Paper 37. Mount Vernon, Ky.: ASPI Publications, 1996.

Reynolds, Michael. *Earthship.* Vol. 1: *How to Build Your Own.* Taos, N.M.: Earthship Biotecture, 1990.

———. *Earthship.* Vol. 2: *Systems and Components.* Taos, N.M.: Earthship Biotecture, 1991.

Roy, Rob. *Earth-Sheltered Houses: How to Build an Affordable Underground Home.* Gabriola Island, B.C.: New Society Publishers, 2006.

———. *Timber Framing for the Rest of Us: A Guide to Contemporary Post and Beam Construction.* Gabriola Island, B.C.: New Society Publishers, 2006.

Sobon, Jack. *Build a Classic Timber-Framed House: Planning and Design, Traditional Materials, Affordable Methods.* Pownal, Vt.: Storey, 1994.

Stanley, Tomm. *Stone House: A Guide to Self-Building with Slipforms.* Christchurch, New Zealand: Stonefield Publishing, 2004.

Stanley Park Earthen Architecture Project. Vancouver, B.C. http://www.stanleyparkecology.ca/programs/cob/research.htm.

Stultz, Roland. *Appropriate Building Materials.* Croton-on-Hudson, N.Y.: Intermediate Technology Publications, 1981.

Vivian, John. *Building Stone Walls.* Charlotte, Vt.: Garden Way Publishing, 1976.

Volunteers in Technical Assistance (VITA). *Making Building Blocks with the Cinva-Ram Block Press.* Arlington, Va.: VITA, 1966.

Wolfskill, Lyle A., Wayne A. Dunlap, and Bob McGalaway. *Handbook for Building Houses of Earth.* Washington, D.C.: U.S. Agency for International Development, n.d.

CHAPTER 22 **CORDWOOD STRUCTURES**

ASPI. *Cordwood Buildings.* Technical Paper 5. Mount Vernon, Ky.: ASPI Publications, 1990.

Roy, Rob. "The Charm of Cordwood Construction." *Mother Earth News,* June/July 2003, 72ff.

———. *Cordwood Buildings: The State of the Art.* Gabriola Island, B.C.: New Society Publishers, 2003.

——. *Cordwood Masonry Houses: A Practical Guide for the Owner-Builder.* New York: Sterling, 1982.

———. *How to Build Log-End Houses.* New York: Drake, 1977.

Snell, Clarke, and Tim Callahan. *Building Green: A Complete How-To Guide to Alternative Building Methods, Earth Plaster, Straw Bale, Cordwood, Cob, Living Roofs.* Asheville, N.C.: Lark Books, 2006.

CHAPTER 23 **YURTS IN APPALACHIA**

ASPI. *The Yurt: An Excellent Low-Cost Home.* Technical Paper 8. Mount Vernon, Ky.: ASPI Publications, 1991.

Easton, Bob. *Shelter.* 2nd ed. Bolinas, Calif.: Shelter Publications, 2000.

Hermann, E. "Yurts!" *Sierra,* November/December 1987, 51ff.

Hunt, Heidi. "Tipis and Yurts." *Mother Earth News,* December 2002 / January 2003, 56 ff.

King, Paul. *The Complete Yurt Handbook.* Bath, UK: Eco-Logic Books, 2002. Distributed by Chelsea Green.

The Mountain Institute (formerly Woodlands Mountain Institute), 1707 L Street, NW, Suite 1030, Washington, D.C. 20036; 202-452-1636; http://www.mountain.org.

Neihardt, John G. *Black Elk Speaks.* New York: MJF Books, 1991.

Pearson, David. *Circle Houses: Yurts, Tipis, and Benders.* White River Junction, Vt.: Chelsea Green, 2001.

The Yurt Foundation, Bucks Harbor, Maine 04618.

CHAPTER 24 **SIMPLE MODES OF TRANSPORTATION**

Alvord, Katie. *Divorce Your Car.* Gabriola Island, B.C.: New Society Publishers, 2001.

ASPI. *ASPI's Solar Charged Electric Car.* Technical Paper 63. Mount Vernon, Ky.: ASPI Publications, 2001.

Backwoods Solar Electric Systems, 1589 Rapid Lightning Creek Road, Sandpoint, Idaho 83864; 208-263-4290; http://www.backwoodssolar.com.

Brant, Bob. *Build Your Own Electric Vehicle.* New York: McGraw-Hill / TAB Electronics, 1993.

Brown, Michael P., with Shari Prange. *Convert It! A Step-by-Step Manual for Converting an Internal Combustion Vehicle to Electric Power.* 2nd ed.

Fort Lauderdale: South Florida Electric Automobile Association, 1993.

Electro Automotive. *Catalog of Components and Complete Kits;* PO Box 1113, West Felton, Calif. 95018-1113; http://www.electroauto.com.

Hackleman, Michael. *The New Electric Vehicles: A Clean and Quiet Revolution.* Ashland, Ore.: Home Power, 1996.

Solatron Technologies, Inc., Supplier of Grid-Connected Solar Equipment, 15663 Village Drive, Suite C, Victorville, Calif. 92394; http://www.partsonsale.com.

CHAPTER 25 **COMPOSTING AND VERMICOMPOSTING**

Appelhof, Mary. *Worms Eat My Garbage: How to Set Up and Maintain a Worm Composting System.* Kalamazoo, Mich.: Flower Press, 1997. Distributed by Chelsea Green.

ASPI. *Composting for Gardens.* Technical Paper 11. Mount Vernon, Ky.: ASPI Publications, 1991.

Ebeling, Eric, ed. *Basic Composting: All the Skills and Tools You Need to Get Started.* Mechanicsburg, Pa.: Stackpole Books, 2003.

Gershuny, Grace, and Deborah L. Martin. *The Rodale Book of Composting: Easy Methods for Every Gardener.* Emmaus, Pa.: Rodale Press, 1992.

Gershuny, Grace, and Joe Smilie. *The Soul of Soil.* White River Junction, Vt.: Chelsea Green, 1996.

Hale, James. "The Earthworm Lawn and Garden." *Mother Earth News,* June/July 2000, 68 ff.

Nancarrow, Loren, and Janet Taylor Hogan. *The Worm Book: The Complete Guide to Worms in Your Garden.* Berkeley, Calif.: Ten Speed Press, 1998.

Noyes, Nick. *Easy Composters You Can Build.* North Adams, Mass.: Storey, 1998.

Sherman, Rhonda. "Backyard Composting Developments." *BioCycle,* January 31, 2005, 45–47.

———. "Vermicomposting Systems Overview." *BioCycle,* December 31, 2002, 53–56.

CHAPTER 26 **COMPOSTING TOILETS**

ASPI. *Compost Toilets.* Technical Paper 2. Mount Vernon, Ky.: ASPI Publications, 1990.

———. *Solving Kentucky's Water Problems.* Technical Paper 41. Mount Vernon, Ky.: ASPI Publications, 1998.

Del Porto, David, and Carol Steinfeld. *Composting Toilet System Book: A Practical Guide to Choosing, Planning, and Maintaining Composting Toilet Systems.* Concord, Mass.: Center for Ecological Pollution Prevention, 2000.

Harper, Peter, and Louise Halestrap. *Lifting the Lid: An Ecological Approach to Toilet Systems.* Machynlleth, Powys, UK: Centre for Alternative Technology, 2001.

Jenkins, Joseph. *The Humanure Handbook: A Guide to Composting Human Manure.* 3rd ed. Grove City, Pa.: Jenkins Publishing, 2005.

Kieffer, Jack. *Preparation of Compost from the Toilet for Use in the Garden.* Technical Paper 41. Mount Vernon, Ky,: ASPI Publications, 1998.

CHAPTER 27 **RECYCLED, SALVAGED, AND DECONSTRUCTED MATERIALS**

ASPI. *Rural Waste Recovery.* Technical Paper 18. Mount Vernon, Ky.: ASPI Publications, 1992.

———. *Wood Waste Utilization.* Technical Paper 47. Mount Vernon, Ky.: ASPI Publications, 1998.

Beard, Daniel Carter. *Shelters, Shacks, and Shanties.* Port Townsend, Wash.: Breakout Productions, 1987. (First published in 1914.)

Broadstreet, Jim. *Building with Junk and Other Good Stuff: A Guide to Home Building and Remodeling Using Recycled Materials.* Port Townsend, Wash.: Loompanics Unlimited, 1990.

Grassroots Recycling Network, 4200 Park Blvd., Suite 290, Oakland, Calif.; 510-531-5523; http://www.grrn.org.

Institute for Local Self Reliance. *Building a Deconstruction Company: A Training Manual for Facilitators and Entrepreneurs.* Washington, D.C.: Institute for Local Self Reliance, 2001.

Selman, Neil. *Waste to Wealth.* Washington, D.C.: Institute for Local Self Reliance, 1985.

CHAPTER 28 PONDS AND AQUACULTURE

Aquaculture Magazine, PO Box 1409, Arden, N.C. 28704; 828-687-0011; http://www.aquaculturemag.com.

Baum, Carl, and Ron Zweig. *Construction of a Solar Algae Pond.* Technical Bulletin 2. East Falmouth, Mass.: New Alchemy Institute, 1982.

Chakroff, Marilyn. *Freshwater Fish Pond Culture and Management.* Collingdale, Pa.: Diane Publishing, 1982.

Fairchild, Bob. *Aquaculture.* Technical Paper 20. Mount Vernon, Ky.: ASPI Publications, 1993.

Lovgren, Gosta H. *The Ponder's Bible.* La Vallette, N.J.: Carolelle Publishing, 2000.

Matson, Tim. *The Country Pond Maker's Guide to Building, Maintenance, and Restoration.* Woodstock, Vt.: Countryman Press, 1991.

Matson, Tim. *Earth Ponds Sourcebook: The Pond Owner's Manual and Resource Guide.* 2nd ed. Woodstock, Vt.: Countryman Press, 2004.

McLarney, William. *The Freshwater Aquaculture Book: A Handbook for Small Scale Fish Culture in North America.* 2nd ed. Point Roberts, Wash.: Hartley & Marks, 1998.

Mims, Steven D. "Aquaculture of Paddlefish in the United States." *Aquatic Living Resources,* vol. 14, 2002, 391–98.

Van Gorder, Steven D. *Home Aquaculture.* Emmaus, Pa.: Rodale Press, 2000.

Wilson, Don. *Small-Scale Crayfish Farming for Food and Profit.* Murphy, N.C.: Atlas Publications, 1990.

CHAPTER 29 CISTERNS AND WATER CATCHMENTS

ASPI. *Cisterns.* Technical Paper 3. Mount Vernon, Ky.: ASPI Publications, 1990.

———. *Water Purification Techniques.* Technical Paper 19. Mount Vernon, Ky.: ASPI Publications, 1993.

Campbell, Stu. *The Home Water Supply: How to Find, Filter, Store, and Conserve It.* North Adams, Mass.: Storey, 1983.

Cordes, Helen. "Cistern Savvy." *Utne Reader,* May/June 1993, 78ff.

Knopper, Melissa. "Drugging Our Water: We Flush It, Then We Drink It." *E Magazine,* January/February 2003, 40.

Ludwig, Art. *Water Storage: Tanks, Cisterns, Aquifers, and Ponds for Domestic Supply, Fire, and Emergency Use. Includes How to Make Ferrocement Water Tanks.* Santa Barbara, Calif.: Oasis Design, 2005.

Nissen-Peterson, Erik. *Rainwater Catchment Systems for Domestic Supply: Design, Construction, and Implementation.* Rugby, UK: ITDG Publishing, 2000.

Pacey, Arnold. *Rainwater Harvesting: The Collection of Rainfall and Runoff in Rural Areas.* Rugby, UK: ITDG Publishing, 1986.

U.S. Environmental Protection Agency. *Manual of Individual Water Supply Systems.* Amsterdam: Fredonia Books, 2001.

Watt, S. B. *Ferrocement Water Tanks and Their Construction.* Rugby, UK: ITDG Publishing, 1978.

CHAPTER 30 **IRRIGATION AND WATER CONSERVATION**

ASPI. *Drip Irrigation Systems.* Technical Paper 25. Mount Vernon, Ky.: ASPI Publications, 1994.

Broner, I. *Drip Irrigation for Home Gardens.* No. 4.702. Fort Collins: Cooperative Extension, Colorado State University, 2004. Available online at http://www.ext.colostate.edu/pubs/garden/04702.html.

Ross, David S. *Trickle Irrigation for Fruit Crops in Maryland.* Agricultural Engineering FACTS 105. College Park: University of Maryland Cooperative Extension Service, 1979.

Sanders, Douglas C. *An Introduction to Drip Irrigation for Vegetables.* Horticultural Information Leaflet No. 33-C. Raleigh: North Carolina Agriculture Extension Service, 1988.

Smith, Stephen W. *Landscape Irrigation: Design and Management.* Hoboken, N.J.: John Wiley & Sons, 1996.

Sneed, Ronald E. *Trickle Irrigation.* NCSU Biological and Agricultural Engineering Water Management Memo. Raleigh: North Carolina Agriculture Extension Service, n.d.

Trimmer, Walter L., and Brian R. Chandler. *Drip Irrigation for Windbreaks.* NebGuide G80-525-A. Lincoln: Cooperative Extension, University of Nebraska, 1981 and 1997.

Yeomans, Ken. *Water for Every Farm: Yeomans Keyline Plan.* Paradise Waters, Queensland, Australia: Keyline Designs, 2002. Available online at http://www.keyline.com/au/form01.htm.

CONCLUSION: AN APPALACHIAN APPROPRIATE TECHNOLOGY

Odum, Eugene P. *Ecology and Our Endangered Life-Support Systems: A Citizen's Guide.* Sunderland, Mass.: Sinauer Publishers, 1993.

Richardson, Jean. *Partnership in Communities: Reweaving the Fabric of Rural America.* Washington, D.C.: Island Press, 2000.

POSTSCRIPT: COMMUNICATIONS

Mander, Jerry. *In the Absence of the Sacred: The Failure of Technology and the Survival of the Indian Nations.* San Francisco: Sierra Club Books, 1991.

Stoll, Clifford. *Silicon Snake Oil.* New York: Anchor, 1996.

Index